POWER OF POSITION

HISTORY AND FOUNDATIONS OF INFORMATION SCIENCE

Edited by Michael Buckland, Jonathan Furner, and Markus Krajewski

Human Information Retrieval by Julian Warner

Good Faith Collaboration: The Culture of Wikipedia by Joseph Michael Reagle Jr.

Paper Machines: About Cards & Catalogs, 1548–1929 by Markus Krajewski, translated by Peter Krapp

Information and Intrigue: From Index Cards to Dewey Decimals to Alger Hiss by Colin B. Burke

Indexing It All: The Subject in the Age of Documentation, Information, and Data by Ronald E. Day

Bibliometrics and Research Evaluation: The Good, the Bad, and the Ugly by Yves Gingras

Search Foundations: Toward a Science of Technology-Mediated Experience by Sachi Arafat and Elham Ashoori

The Information Manifold: Why Computers Can't Solve Algorithmic Bias and Fake News by Antonio Badia

Documentarity: Evidence, Ontology, and Inscription by Ronald E. Day

The Infographic: A History of Data Graphics in News and Communications by Murray Dick

The Typographic Medium by Kate Brideau

Power of Position: Classification and the Biodiversity Sciences by Robert D. Montoya

POWER OF POSITION

Classification and the Biodiversity Sciences

ROBERT D. MONTOYA

The MIT Press
Cambridge, Massachusetts
London, England

This book is freely available in an open access edition thanks to TOME (Toward an Open Monograph Ecosystem)—a collaboration of the Association of American Universities, the Association of University Presses, and the Association of Research Libraries—and the generous support of Arcadia, a charitable fund of Lisbet Rausing and Peter Baldwin, and the UCLA Library. Learn more at the TOME website, available at: openmonographs.org.

The MIT Press would like to thank the anonymous peer reviewers who provided comments on drafts of this book. The generous work of academic experts is essential for establishing the authority and quality of our publications. We acknowledge with gratitude the contributions of these otherwise uncredited readers.

This book was set in Adobe Garamond Pro by Westchester Publishing Services. Printed and bound in the United States of America.

Library of Congress Cataloging-in-Publication Data

Names: Montoya, Robert D., author.
Title: Power of position : classification and the biodiversity sciences / Robert D. Montoya.
Description: Cambridge, Massachusetts : The MIT Press, [2022] | Series: History and foundations of information science | Includes bibliographical references and index.
Identifiers: LCCN 2021033972 | ISBN 9780262045278 (paperback)
Subjects: LCSH: Biology—Classification. | Life sciences—Classification. | Cladistic analysis.
Classification: LCC QH83 .M68 2022 | DDC 570.1/2—dc23/eng/20211221
LC record available at https://lccn.loc.gov/2021033972

10 9 8 7 6 5 4 3 2 1

Simply, to Baca

Contents

Acknowledgments

I would like to thank so many people here, for this book represents the accumulation of so many conversations, papers, courses, panels, happy hours, Zoom sessions (ugh, pandemic!), and late-night chats. I am very grateful to my editor at the MIT Press, Gita Manaktala, for her support and kindness throughout this process. Much of this book was written in the throes of a global pandemic and a cross-country move, and it was Gita's patience and willingness to offer more writing time that allowed me the mental space to complete the manuscript. I thank Erika Barrios, also at the MIT Press. I, of course, thank the anonymous peer reviewers for their investment in my work and their wonderful and extensive feedback. While the manuscript has transformed in whole, the genesis of this manuscript was my PhD dissertation at the UCLA Department of Information Studies. The research for this book was supported by a generous Doctoral Dissertation Improvement Grant from the National Science Foundation. I am indebted to my committee, Johanna Drucker, Jonathan Furner, Geoffrey Bowker, and Chris Kelty. Beyond them, I am grateful for the hours and hours of conversation with some of the brightest colleagues I could ever have the pleasure of working with: Gregory H. Leazer, Ronald E. Day, Joseph T. Tennis, Michael Buckland, Sarah Roberts, Safiya Noble, Michelle Caswell, and Howard Rosenbaum. I am a lucky person, indeed.

This book could not have been written without the patience of colleagues involved in the various biodiversity and natural history organizations I discuss: Thomas Orrell of the Smithsonian National Museum of Natural History, ITIS, and the Catalogue of Life; Yury Roskov of the Catalogue of Life; Donald Hobern, then executive director of GBIF; Peter Schalk of the Naturalis Biodiversity Center and Catalogue of Life; Rebecca Snyder of the Smithsonian National Museum of Natural History; Holly Little of the Smithsonian National Museum of Natural History; Tim Robertson of GBIF; Paul Kirk of Royal Botanical Gardens, Kew; Paul Kirk of the Royal Botanical Gardens, Kew; Timothy M. A. Utteridge of the Royal Botanical Gardens, Kew; Alan Paton of the Royal Botanical Gardens, Kew; and Vince Smith at the Natural History Museum, London. Academia is tough space, but the generosity of my colleagues who worked with me through this project is beyond reproach. I thank the many students I've learned from over the years. I thank Steve Barnthouse, Julie Van Winkle, and Rahul Subramanian. And finally, as a first-generation Latino college student, I thank my family: in particular, my mother, Elena Montoya, for working so hard to make sure I had the luxury of pursing my educational career, and my uncle, Alfred (Baca) Delgado, who instilled in me the love of learning about botany, zoology, and so many other things.

INTRODUCTION

AN IMPERATIVE AND INITIATIVES

In June of 1992, a meeting of nations, collectively called the Earth Summit, gathered at the United Nations Conference on Environment and Development in Rio de Janeiro, Brazil, to discuss pressing issues facing the sustainability of biological life on the planet. An outcome of this meeting was the adoption of Agenda 21, a "wide-ranging blueprint for action to achieve sustainable development worldwide" (United Nations 1997, 2017). Topics at the summit included the lack of access to potable water across the globe, the increasing use of fossil fuels, the expanding production of toxic waste, and the decline of earth's biodiversity due, in large part, to the detrimental influence of human activities on global ecosystems (United Nations 1997). An important international treaty arose from this summit, the Convention on Biological Diversity (CBD), which articulates the critical need for coordinated scientific information-exchange infrastructures to better understand and reverse the globe's diminishing biological diversity (2016). A core focus of the CBD is the general acknowledgment that the fruits of biodiversity research (data, nomenclature, species lists, taxonomies, and the like) are much more than "merely" the identification of "plants, animals and micro-organisms and their ecosystems" (Convention on Biological Diversity 2017b). This knowledge also has radiant, derivative influence across many disciplines, and thus has an increasing pertinence to, and impact on, global populations in all corners of the planet (*all* biological populations, including, but not limited, to humans).

The historical impact of this document has been significant, for it serves as the motivating and pro forma international agreement on which numerous biodiversity and ecological initiatives have built endorsement for their respective projects. On a local level, the CBD has spurred the implementation of laws and policies that govern a number of domains pertaining to biodiversity issues (Kate 2002). The CBD has arisen at a watershed moment in biodiversity-related data practices, in particular, and has acted as a catalyst for the production of numerous centralized data repositories intended to bring together localized knowledge sets within openly accessible, global infrastructures to facilitate information exchange. In particular, Article 7 of the CBD explicitly acknowledges the importance of maintaining and organizing data derived from the identification and monitoring of biological diversity (1992, 5).

Building on this acknowledgment, during the 1998 Conference of the Parties—the governing body of the convention—national participants articulated the Global Taxonomy Initiative (GTI) to address what they referred to as a prevailing "taxonomic impediment" (Convention on Biological Diversity 2017a; Hopkins and Freckleton 2002). The impediment cites a "shortage of taxonomic expertise, taxonomic collections, field guides, and other identification aids," caused, in part, by the general difficulty in accessing existing taxonomic information (Convention on Biological Diversity 2003, vii). According to the CBD, the GTI marks "the first time in history that taxonomy has had recognition at such a high level in international policy" (2003). Given the CBD's articulated concern with the fragmented nature of biodiversity knowledge, the GTI articulates clear steps by which authoritative online platforms should collocate regional taxonomic information by strengthening "regional cooperation" (2003, 1). Locally specific biodiversity knowledge has historically been stored in site-specific ways throughout the globe, effectively unavailable to the larger scientific population for integrative work. Researchers have also generally been (and continue to be) trepidatious, flatly unwilling, or technically unable to share their data. Yet, given the reality that grant-based and governmental funding is limited and in high demand, centralizing data in large databases—and thereby centralizing activities such as species naming and describing and taxonomy building—is one approach scientists can take, collectively, to sustain a long durée approach to biological taxonomy (Ribes and Finholt 2009; Thomas 2009). Thus, in response to the Global Taxonomy Initiative's call, large-scale federated platforms have begun to gain operational steam, aggregating taxonomic and descriptive data with the eventual goal of unifying all geographically specific caches of biodiversity knowledge (Bowker 2008, 120; Waterton, Ellis, and Wynne 2013, 108).

Despite the existence of initiatives such as the GTI, and the multitude of newer initiatives that have arisen since the initial 1992 gathering, global biodiversity continues to decline, in part because of large-scale global phenomena such as climate change and global warming, suggesting that there is far more work to be done (Bellard et al. 2012; Butchart et al. 2010). Gaining a more robust understanding of the biodiversity of the planet is a pressing scientific task if we are to understand the full impact of this new ecological reality. However, the ability for scientists to study and understand the scope of ecological issues—especially global ones—rests on the scientific community's capacity to name, document, and classify, and then *communicate*, our collective knowledge about what species exist on the planet and how these species are being affected by these new conditions. As we will learn, aggregating species information is no easy task, nor is aggregating this information into *normalized* systems across numerous research teams. This normalization of data format presents various structural and epistemic problems; by taking a global view of data, we can begin to consider how we might solve problems that exist on a global scale.

Thankfully, many recent scholars have identified the importance of understanding the operationally and technologically focused aspects of the biodiversity sciences. Each of these scholars notes that, if we are to aggregate appropriately, and with respect to the constitution of scientific knowledge, we must also attend to the data practices and resultant social epistemological impacts of collective taxonomy production. This increasing focus on data and information in the domain has been described by Catherine Kendig and Joeri Witteveen as a burgeoning "information science turn" in taxonomic studies (2020). In a comprehensive special issue of *History and Philosophy of the Life Sciences*, Kendig and Witteveen bring together a group of scholars to examine how taxonomic practices (vis-à-vis information control) intersect with the scientific procedures of communication, naming, and taxonomy. So, while the practice of aggregating data within the biodiversity sciences may seem innocuous and somewhat neutral to the general reader, the reality is that this work has significant impacts on taxonomy as it intersects with the domains of ontology, epistemology, and, as we will find, on the circulation of power and control within these classification systems. This book follows a long line of empirically informed research such as Sabina Leonelli's, *Data-Centric Biology: A Philosophical Study* (2016), and Christine Hine's early study, *Systematics as Cyberscience: Computers, Change, and Continuity in Science* (2008). What both of these scholars illustrate is that within the fields of systematics, biology, and taxonomy, the practices of organizing and building systems of knowledge—"making science happen," so to speak—need to be attended to if we are to understand the rapidly increasing effects of technology on the constitution of scientific ideas as a historically contextualized, socially affective, and epistemologically nuanced phenomenon.

The problematics of coordinating and producing such taxonomic infrastructure for biodiversity, however, are numerous and complex, particularly if we look more closely at the specific technological databases that have arisen to aggregate classification and taxonomic information. In the last twenty years—spurred on by, if not created as a direct result of, the CBD's articulated aims and directives—new federated digital initiatives such as the Global Biodiversity Information Facility (GBIF) (2017) and the International Barcode of Life (2015; Waterton, Ellis, and Wynne 2013) have taken on the management of worldwide biodiversity data toward the end of universal and standardized access. These and many other information systems are collectively used in scientific research to direct global biodiversity initiatives supported by governmental and nongovernmental organizations toward the development of policies, as well as to make decisions for issues as far reaching as climate change research, global health initiatives, and conservation assessments (Jetz, McPherson, and Guralnick 2012, 151). Yet, as our contemporary political period has soberingly illustrated, while classification data are used as decision-making variables in public discourse, we cannot assume that the nonspecialists and governmental officials who use them as tools for the articulation of policy and law, for example, will have the capacity—or frankly, the ethical principles—to understand what these classifications can and cannot tell them and to invoke them responsibly.

This is the gist of the problem this book hopes to address: that classifications are powerful mechanisms in our information-rich world and that we must better attend to the machinations inherent in this power, as well as to how the effects of this power proliferate beyond the boundaries of their original intent. I emphasize power as a core analytic because classifications are nothing if not systems that relate one entity with another—a task that inherently requires individuals (or, more detrimentally, algorithms) to value an entity's relative importance (by way of its position) within a system of other entities. Classifications involve two spaces of concern: an internal, representational space that depicts species relationships (a classification of canids, for example), and a material space that produces effects on external bodies, such as natural organisms, people, and communities, in that how it represents them impacts their identity and possibilities in the external world. For example, classifications have a great deal to say about how much power society, or a government, might or might not have over nature. Classifying the Great Green Macaw (*Ara ambiguus*) as an endangered species in Costa Rica, as a case in point, has made it a species worthy of conservation, thereby (potentially) offsetting the rapidly dwindling numbers of birds in the Monteverde Cloud Forest. The same can be said of the contested dingo of Australia, though the jury is still out on whether it deserves protection as a species or whether it should be eradicated as a pest. Shall we continue to kill dingoes on account that they are an ecological pest? To some, perhaps. Such decisions partly depend on what their scientific name is (and by extension, what taxon concept it represents)—on their *position* within some reference classification. For obvious reasons, these results can be destructive to our ecosystems, but these classifications can also, in a roundabout way, be a mechanism that provides agency to a species that would otherwise have no legal or cultural protections over its being. Organizations like the International Rhino Foundation and the World Wildlife Fund work daily to bring attention to species that need a voice in a global space that views species as commodity, capital, and game.

Inverting this emphasis, classifications can equally be said to have power over us (humans) and direct our actions in relation to nature. As in the examples above, they most obviously tell us what we can and cannot do with respect to certain species. Classifications either restrict our agency over nature or increase our capability to extract value from it. On top of this, however, we also see *through* classifications, often darkly. They help define nature for us by way of visual and representational cues, but the terms of those definitions may be unclear or impossible to unpack. Nature, after all—whatever this term might mean to any individual—is a complex concept, full of interrelated organisms that are not easy to categorize into neat and discreet species and taxon groups. Nature is also a space to immerse ourselves into and enjoy, but it is also a resource that must be exploited to survive. The extent of this exploitation exceeds what we need for mere survival, of course, since our global population has, in no uncertain terms, made a disaster of our biological ecosystems in exchange for luxury, mobility, and expansion.

More subtly, biodiversity classifications have an innately epistemic power over us. They position our own sense of identity in relation to *the natural.* That relationship is often anthropocentric, as can be seen in the long history of the natural sciences. The Great Chain of Being had human—often, more specifically, "l'homme"—categorized just below the divine. Animals, in Aristotle's view, were to be ranked above plants, given that they sense, breathe, and move. And downward the hierarchy proceeds through quadrupeds, birds, lichens, rocks, minerals, and so on. Biodiversity classifications narrativize nature and give us entree into its complexity by way of simplification and individuation. If everything was not "named" (as a species, a plant, an animal), then we could not communicate our relation to it. It was thus why Hope Olson (2002) was able to declare that power is in a name—that the act of naming itself creates entities that can either be foregrounded in our imagination, pushed to the background of our thoughts, or forgotten altogether within a system of other names that, by virtue of their appearance, are more valued than those which we (actively or passively) choose to obfuscate. To say that classifications *merely* mediate between information and a broader public (scientists, policy makers, conservationists, and the like) is to hopelessly fall into an antiquated information-as-conduit discourse. Classification systems are more than a "service" for a user. The relationship between a system of organization and its user is reciprocal: we may produce classifications, but those classifications also affect our ability to make sense of the information contained within it. In describing the user's relationship to information, Ron Day posits "a model that views subjects and objects as co-emergences mediated through co-determining, contextual (or 'structural') affordances and through in-common zones of mutual affects" (2011). The subject (the viewer of the classification) and the object (the classification itself) are linked in complex and processual ways.

SOCIONATURALITY

The reality that organisms live out in nature as pure and clear-cut named individual taxon groups is a fiction, and one well-known by the scientists charged with naming and classifying species taxa and other taxon ranks. Things do not emerge with names—it is our language and our need to communicate ideas that require us to go about this operation of naming. Nor is the idea that humans—positioned as we are in our own infinitesimally small spot on the tree of life—stand apart from nature in some way a natural circumstance. On one hand, this seems like a very obvious point, yet on the other, most individuals navigate our information ecosystems as if the categories, labels, names, and information structures they navigate are a given. Google results are seldom questioned by the general public, for example. As Judith Butler notes in *The Force of Non-Violence* (2020), figures such as Hobbes and Rousseau present a history of the world that begins with the individual. To illustrate this, Butler presents the fiction of "Robinson Crusoe, alone on an island, providing for their own sustenance, living without dependency on others, without systems of labor, and without any common organization of political and economic life" (2020, 31).

The story of history (and of political economy, for that matter) just kind of "starts there," as a man on an island struggling to fight the onslaught of the natural world. That is, until other individuals arise whose interests conflict with his (and it always is "his") "individual" nature. Society then grows more complex, and individuals must engage in methods of economic exchange, and they are forced to pit their interest in relation to, and against, all others. Social and racial contracts arise, intentionally tempering conflicts through tacit agreements, monitored by our own adherence to the contract or by way of social norms, law, policy, and so on (Rawls 1999; Mills 2011). Yet the true reality, as Butler notes, is that we are nothing if not born into a world of dependency, both as a child and throughout our life as we navigate an environment that, through no deliberate choice of our own, we depend on to subsist through its myriad resources (for food, for shelter, and so on). When we are born, we enter into a world that is not, as we often assume, naturally divided into two components: one of "society" and another of "nature." Rather, we emerge into a reality that is socionaturally entwined into an endless net of what Butler calls "radical dependencies" (2020, 41). It is from this perspective that this book's narrative emerges—that the process of classification and representation broadly separates, rather than connects, our selves with regard to species.

Attending to this assumed bifurcation between nature and human, I think, is particularly important in this moment in history that, if we want to be generous, is at best tumultuous, or at worst, apocalyptic. Some have called this era of systemic change the Anthropocene—a moment when the world has been so drastically altered by humanity's impacts that we have altered the very systems that have steadily maintained life on earth for millions of years. I can imagine no entity more vulnerable at the moment than the natural world and the cycles that define it. Ecosystems are out of balance and global warming continues to worsen, even while governments ignore its existence. At the moment of this writing, each year seems to bring a new record-setting march of hurricanes across the Atlantic Ocean. Glaciers are detaching from permafrost areas at a rate too astonishing to comfortably acknowledge. In one season, the ecosystems of Australia are destroyed by wildfire, while in another it is Greece or Brazil; I am currently writing in the thick, unbreathable air caused by fires on the western coast of the United States. The occurrence of the next catastrophic fire is not a question of if, but when. There is no end in sight.

Others have maintained that to merely declare these impacts to be human driven overlooks the fact that not all humans have contributed to this disaster evenly. "Capitalism, not industrialization," as is commonly articulated, "caused the Earth's transformation by producing massive social inequalities that supported 'audacious strategies of global consequent, endless commodification, and relentless rationalization'"—a process termed the Capitalocene (Ellis 2018, 136; Moore 2016, chap. 1). This is to say that the effects of the Anthropocene, Capitalocene, or whatever other term you might adopt, are unavoidably political. Any new era (or "—cene," as the case may be) is defined by massive shifts and radical biological changes, such as the massive biodiversity loss and extinction events that we are experiencing at the current moment (Moore 2016, 21).

Were it not for the work of biodiversity scientists, and the species lists and classifications they carefully produce, we would have no measure of this loss, nor a language with which to name what we lose wholesale in the process of this change. And, if one of the capacities of biodiversity classification is to help us imagine what nature is, it can then be a prime site wherein humanity can reimagine what it means to be human in a more intimate and ecologically focused way. Now, more than any other time in history, naming species is of the utmost importance, if only so that we can mourn their disappearance and force ourselves to see the direct and specific human effects of our unjust actions toward our environment. As appropriately stated by Manuel Arias-Maldonado, "Of course, animals occupy a key position in the human-natural relation. Actually, they perform an important propagandistic function, in that certain animals—aptly called 'charismatic' ones—symbolize the plight that all of them suffer under human dominion" (2015, 106). Nomenclature and classifications have a heavy responsibility in this tumultuous time by representing organisms both big and small for the equal and distinct value they hold in our ecologies.

My intent here is not to descend into the depths of environmental despair, easy as that may be these days, but I do want to express that the presentation of biodiversity information, especially in the guise of graphical classifications, has a large role to play in this time of ecological change. As stated above, centuries of classification work have held humans to be the apex of the natural world, and this deep-seated cultural belief is still alive and well, even while our scientific classifications change to represent more connectivity. So, while this naming of species is necessary, as is the construction of the classification systems that express them, these processes also simultaneously distance us from nature. Nature is delivered in neat packets that summarize the state of nature as if it is complete and naturally ordered. These technologies help us "exert control over nature, complicating the socionatural relation at the same time, while also being key . . . for redefining that relation in a sustainable or even more caring way" (Arias-Maldonado 2015). Whether these classifications are used to control and exploit or to nurture and heal is based, in a large part, on the way we present their content as either definitive and distant or fluid and relational.

What I mean by this is that classification builders can and should collectively work to emphasize ecological points of view, over views that might otherwise make nature seem like the "other." As also noted by Arias-Maldonado (2015, 8–9), the study of "nature" proper has become multidisciplinary, studied in areas far outside the boundaries of the natural sciences, from the environmental humanities to sociology, economics, and information science. Nature as an entity of study is not only the domain of the empirical sciences, in part because other areas of study note how nature has a part to play in each aspect of our lived experience: in our imagination, in our social circumstances, in the production of our social inequalities, and in the facilitation of our individual powers and abilities to improve our position in society. To understand the broad impacts of biodiversity classifications, we must also understand that their influence is not only representational and inert; they are also complex and nuanced entities that are built to do both practical and epistemic work in the world—quite apart from the fact that we may or may not fully intend for them to be used in one way or another.

Woven into the narrative of this book, I make the case for a few theses:

- Biodiversity classifications, and the entities they organize, are constructed, artificial entities, yet they are popularly often seen as given and "correct." This is not to say that they are scientifically invalid, but it does mean that their epistemic realities are built and not given.
- The natural world is process-oriented, and how we identify entities within it (taxa, species, and the like) will differ depending on our epistemic orientation, which poses challenges in spaces such as in the case of the Catalogue of Life.
- Biodiversity classifications have epistemic and material impacts in the world that radically impact individual and collective being (human and nonhuman alike).
- Universal biological classifications are a detriment to the future of our knowledge space.
- Biodiversity classifications, thus, have a role to play with regard to ecological and environmental justice.

The social powers that classifications exhibit in the lived, social world are ontological, organizational, epistemic, and historically situated. To better understand this power, we must idealize ways to frame how this power is exerted in concrete and manufactured ways. To this end, classifiers have a certain obligation to express how these classifications fit within the broader field of social use and the contexts within which they will function as tools of decision making. So too do the users of these systems, who need to unpack the nuances of their construction.

This narrative, then, is just as much about the social impacts of biodiversity classifications. The overarching argument of this text is that, first, classification systems, in general, are instruments of power, and that, second, within the biodiversity world, this means that biological taxonomies are inherently caught up in the practical and political work of quantifying nature, which influences how the scientific and social world envisions, questions, contains, and liberates the natural entities that we identify in the world. The aim is to better understand how contemporary modes of data aggregation—a relatively new phenomenon in the biodiversity sciences—influence the production of scientific activity as well as the interpretation of the natural world to members of the public, who are increasingly using these interfaces to access data on the natural world.

The above theses can be tweaked to map how these ideas can be generalized to refer to all classifications, and how they function in social space.

- Classifications, and the entities they organize, are constructed, artificial entities. We can have many multiple valid classifications functioning at one time.
- The social world is complex, and to attempt to reproduce its identity in systems will implicitly reduce the world to match the point of view and social assumptions of its builder.
- Classifications have epistemic and material affects in the world that radically impact how individuals can negotiate their being in society.
- Universal classifications are a detriment to the future of our knowledge space, even considering their parsimonious and standards-based approach.
- Classifications, thus, have a role to play with regard to social, cultural, and epistemic justice.

Extending this argument, my goal is to emphasize the extent to which these classifications are constructs that could have been otherwise, had we approached any one of them with a different mechanism for quantifying entities (by size, by shape, and so on) and a different theory for articulating their boundaries. Which is also to say that the "species" that we recognize within classification systems are not "real" or given categories in the natural world: there may be an *entity* we call the Great Green Macaw, but that entity might have previously been identified as some other species. And were we to take a different epistemic point of view, we might even more strongly associate that Great Green Macaw as less a species and, more importantly, as a spiritual entity, or an entity that protects the air and winds, as is the case in some indigenous tribes. What I call *derivative positionality* is critical here: how an entity is positioned within a classification (derivatively) has powerful implications over how that entity is positioned in the lived, material, and social world. And one way to offset and regain control over this social and material power is to expose how that classificatory position is not, in fact, fixed but can be conceptualized in many different capacities in different epistemic contexts. For every classification that we encounter, an equally valid classification could present itself.

Finally, I explore how we can think about complexity and plurality in these spaces. If classificatory positions are not assumed to be fixed, given that one entity might inhabit many different positions in different classifications schemes, we need to propose a mechanism by which we can expose this fluidity. We should work toward creating classifications that also allow for the radical possibility of upsetting that assumed order and imagining new arrangements such that socionaturality and fluid boundaries are exposed rather than obfuscated. Positing pluriversality is, in part, to imagine otherwise, and so we must attend to how we can connect the given with the absent, and the expected with the unexpected. And so, if the process of classifying is disentangling nature into its constituent parts and ordering it in ways that make sense to our epistemic understanding of the world, as affirmed by Butler (2020), what does it mean to *re-entangle* nature to imagine new possibilities that illustrate a dependency-forward way of perceiving entities?

A DERIVATIVE APPROACH: COMPOSITE CLASSIFICATIONS

The central portion of this book examines the case of *composite taxonomies* as a springboard from which we can concretely understand the epistemic limits and potentials of classifications. Composite taxonomies, to state it somewhat reductively, are understood in this project as *derivative taxonomic arrangements that aggregate multiple, subsidiary taxonomies into one universalizing space*. The Catalogue of Life (abbreviated henceforth as CoL or the Catalogue) will serve as the primary example of this type of classification. The Catalogue asserts itself to be an authoritative, management-oriented biodiversity schema that is designed to serve two functions: to provide (1) a single, integrated, and validated species checklist, and (2) a management hierarchy (classification and taxonomy) that can bring together data from different sources representing different taxonomic commitments into one hierarchical design (Species 2000 2017a). Any and all classifications are complex in their own right, but entities like the Catalogue are especially so, since they bring together potentially conflicting schemas that must be edited to the organizational, data-management-oriented commitments of the Catalogue's space.

Aggregating taxonomies in this fashion is not apolitical work, nor is this approach universally recognized by the broader taxonomic community as an effective means of aggregating taxonomic data. Conflicts arise as practical and pragmatic approaches to data collection and collocation are positioned in tension with the hermeneutic and hypothesis-driven work of scientific taxonomic production. The former obfuscates and confuses the empirical work of the latter. If you are a trained biodiversity scientist, such an editorial approach might trigger a (quite reasonable) body-cringing response. Each taxonomy, generated by every scientist or team, is produced under a certain set of intellectual conditions: assumptions about what constitutes a taxon concept, as well as the engrained suppositions about how these concepts should be related based on any number of morphological, genetic, or ecological traits. These assumptions are both metaphysical and ontological (as in, what *kinds* of things exist in the world, and how do they relate), as well as epistemic (in that the classifications are constructed under assumptions about what represents a true and valid representation of the natural world). Systems of any kind, and taxonomic classifications no less, are contingent historical reconciliations, based on current and present knowledge sets that maintain an equal footing in the laboratories "of the past" (Rheinberger 2010, 89–90). Pick a taxonomy—any taxonomy—and you will find a network of knowledge that represents years and perhaps decades of layered and accumulated information, research, and hypothesis formulations.

Despite this unavoidable and complicated reality, the Catalogue has taken upon itself the commingling of these diverse and multiple taxonomic constructions into one unified space. Yet, the Catalogue's stance is that *information must be shared* in order for biodiversity knowledge to reach its full research impact and potential. And to reach this potential, standards need to be implemented, even if contributed taxonomies must be manipulated to cohere with global data standards. Concerned as the information studies (IS) community is with pluralistic approaches to classification

(Mai 2011; Szostak 2015) and the representation of diverse voices and fluid ontologies in and for our information systems (Seddon and Srinivasan 2014; Srinivasan et al. 2009; Srinivasan and Huang 2005; Srinivasan, Pepe, and Rodriguez 2009), spaces such as those inhabited by the Catalogue can be incredibly instructive toward these just ends, if nothing else as provocative starting points for discussions about what plurality should and can look like in practice.

But this is just one part of the story: as a management classification, the Catalogue is also integrated into other systems as core organizational architecture. Once the Catalogue is compiled, it is subsequently embedded into a network of other biodiversity systems, thereby amplifying its effect across the landscape of biodiversity practice. And given that contemporary database taxonomies are now the main source of taxonomic knowledge, the consequences of this activity can be globally consequential (Hodkinson 2011; Parr et al. 2004; Watson, Lyal, and Pendry 2015, chaps. 2, 9).

DISCIPLINARY FRAMING AND THE BROADER CONCERN

A note must be included about the disciplinary gaze of this book—that it is derived from the field of information studies is significant. "The world is full of writings," wrote IS scholar Patrick Wilson, "Most are only of passing interest to anyone, despite their being records or traces of human activity; not all of our history is worth remembering" (1968, 1). While often flying fairly low under the academic radar, the domain of IS has been of great consequence to each and every facet of scholarly production. Any scholar working within an academic setting since, conservatively, the late nineteenth century has been implicitly producing scholarship with the books, data, documents, and objects that librarians, archivists, data experts, and museum curators have found *fitting* to preserve as a representation of human activity. This statement may seem bold, but to my mind, its heft is warranted and unequivocally true. For every "discovery" made within the limits of an archive, museum, or library, there is a long line of individuals, including a librarian, archivist, or museum curator, who *chose* to save an item for posterity.

One might recall the repartee between the historian of science and technology Suzanne Fischer (2012) and freelance researcher and former librarian Helena Iles Papaioannou (2012) regarding a Lincoln report found in the US National Archives. At the center of this argument was a medical report on Abraham Lincoln sent to the then US surgeon general by the first doctor to arrive at Ford's Theatre after the president was shot by John Wilkes Booth. The report was, as you might imagine, housed in the surgeon general's records at the National Archives, filed under "L," for the name of the doctor, Charles Leale. News reports of this (no doubt) important document touted the letter as being "discovered," "unearthed," "rediscovered," and "found." Fischer's opinion was that no document is "discovered," and that it was just where it should be: where an archivist had put it. Papaioannou then responded, claiming that no archivist knew of the report and that its existence was, indeed, unknown, making discovery possible. "The title of [Fischer's] article ["Nota Bene: If You 'Discover' Something in an Archive, It Is Not a Discovery"] suggests it is impossible for a researcher to make an archival discovery," Papaioannou wrote (2012).

In most cases, I'd probably remain neutral on this argument, given the somewhat tedious nature of arguments of this kind, but there is no scenario in which "discovery" is an appropriate term. As a librarian who has worked in academic libraries and archives most of my career, I know the energy it takes to convince countless researchers of the worth of information professionals—a Sisyphean feat. If anything is to be discovered, it should be the institutional arrangements of power that facilitate the acquisition, appraisal, maintenance, and preservation of collections. What researchers do accomplish, of course, is articulate the social worth of documents within their discipline at particular points in time. That Papaioannou brought the letter to the public as part of a research endeavor is incredibly important: the document did, indeed, shed new information on a matter of immense historical import. This contribution should not be discounted or undervalued, and I hope that is clear to all readers, including those involved in this controversy. The nuance in Papaioannou's argument is that the letter was not catalogued at the item level, nor was there any documentation regarding the decision-making process that led to keeping *that individual letter*. Given this, it is unlikely that an "appraisal decision was made on any particular document" within the surgeon general's records (Papaioannou 2012).

The reality, however, is that just because a process is not documented does not, by default, mean a process of selection did not take place. The document is *there*: it is in the archive, and that should be documentation enough to support that the item was, in fact, curated to be just where it is. To assume this to be the case is the professional courtesy I give to those who worked hard to preserve it. Surely, an archivist may not know of the true value (if such a concept exists) of a document at a given point in time, within a certain discipline, but that does not stop them from performing due diligence to foresee what might be important to current and future scholars. Archivists are not prescient, and certainly are not experts in all fields. They value collections by using context to the best of their abilities. In this case, the surgeon general is an important historical figure, and thus, their collection typically merits archiving. Archivists are professionals trained to manufacture clarity from within the countless number of documents produced during the course of daily activity. As Wilson said, "Not all of our history is worth remembering." Information professionals, nevertheless, take it upon themselves to manage the impossible task of crafting cultural memory and assigning value to the voices of some and not others. It is not easy, but it must be done.

I begin this section on disciplinary framing with this story because it shows why IS is such an important, if undervalued, discipline in the academy. It doesn't matter whether we are talking about a document in the archives of the National Library or a type specimen of a biological species from the Central American tropics in the London Museum of Natural History. There is a story behind each concerted decision to collect each and every object, and this decision-making process matters when we think about the scholars who generate scholarship by using these collections. Thus, the power of what can and cannot be integrated into ongoing scholarship is an unavoidable reality of information work. Librarians, archivists, and curators are poised daily to make decisions about what other evidence disciplines will use as the basis for their ongoing theoretical and methodological investigations.

But IS is more than the institutional activities that we recognize as information work. The discipline also brings with it a rich and critical mode of engagement with information that merits broader appeal in adjoining disciplines. IS a broad term that, in the context of this book, is inclusive of information science, an umbrella term covering many of the data-oriented domains that have arisen in recent years, such as informatics and data science. The use of the term "studies" in information studies is vital, because my approach is far more humanistic and cultural in its constitution than one might otherwise assume when thinking about "information science" proper. And certainly, if one looks to the content of data science programs—seemingly ubiquitous, popping up in departments ranging from business and economics to statistics and artificial intelligence—what is often lacking in them is a critical focus. By critical I mean an approach that examines objects of interest with an eye toward embedded power structures, ethical possibilities, and just ends. Surely there are exceptions, but these exceptions do not define the business-driven Silicon Valley mentality that pervades the typical approach to "data" in these many domains. And this mentality is dangerous to our health as a collective community.

The field of library studies is important to highlight here as well, especially since social justice and activist-oriented values have long been a central part of the profession. The American Library Association holds values such as "access, confidentiality/privacy, democracy, diversity, education and lifelong learning, intellectual freedom, preservation, the public good, professionalism, service, social responsibility, and sustainability" as core values to librarianship. In library studies we strive to uphold one's individual right to epistemic freedoms, emphasize community experiences and concerns, and prioritize globally comprehensive notions of truth and justice.

A Comparative Approach

My approach in this book is comparative, in the sense that I am looking to biological classification in hopes of understanding some general epistemic qualities about what it means to classify *at all* in our contemporary world, and the effects these qualities have on social spaces (Danton 1973). This approach is obvious in the narrative, in that I bring to bear examples from both the biodiversity world and the "traditional" library world. But to be clear: my goal here is not to postulate a "general system" of classification—in that my examinations should somehow hint at some fundamental, normative techniques—nor do I claim that there are universal ways in which to understand classifications. To do so would be antithetical to the purpose of this book. However, I do think that the differences between library and biodiversity classifications, in terms of both theory and practice, can help us envision solutions that best situate classifications with more pluralistic capacities. The reality is that what constitutes scientific knowledge has always been contested. In the biodiversity sciences, some groups hold epistemic values that are in contradiction to other groups. This is the sign of a healthy discipline, focused on pushing the limits of our understanding of nature.

The comparative approach is also vital if the fields of classification studies and IS are to better understand the lasting and often-repressive effects of major epistemic cultural shifts such as colonialization, capitalism, and globalization. In this sense, I see the comparative method as a core approach to a critical study of representation systems, insofar as the method can tease out how super-structural modes of power (culture, politics, law, education, and so on) express themselves in the domain of librarianship by way of organizational, descriptive, collection, and access practices mediated through various technological and epistemic instruments. It is useful, as well, to exit our spaces of disciplinary comfort to find surprises in these distinct domain specificities. My argument here is that we need more of this work, for the situated, contextually specific, and historically contingent attributes of these knowledge systems can inform, broaden, and render more pluralistic our understanding of classification and knowledge organization in general. Such work, I believe, is vital to the longevity of the discipline of IS.

There is also a historical precedent for this approach. As Ronald E. Day states, "Universal bibliographical classifications and descriptions followed the example of zoological taxonomy and classification in the century before them" (2014, 39). Seen in this light, this project seeks to return back to these roots, to reengage IS scholarship in the organizing endeavors and practices of the natural sciences. Taking a close look at the practices of biodiversity scientists in relation to practices in information studies isn't an altogether strange juxtaposition, for, as David Hull states, "as most people view taxonomists, they are more librarians than scientists and just as loveable. . . . Collectors and classifiers were the ones who had sufficient knowledge to appreciate the true diversity of life" (1988, 81).

In IS, our aim might be to organize books, documents, or data, but the classificatory assumptions we use for any particular system are anchored only by the artificial boundaries and suppositions imposed by the classifier. The Library of Congress, for example, is organized by discipline. Class B contains works on philosophy, psychology, and religion, while Class Q contains documents related to science. Books and documents are not created out of a naturally occurring system or ecology to ascertain the extent or boundaries of our classificatory possibilities. The artificiality of classificatory systems is, in part, what drove Hope Olson, in her influential text, *The Power to Name: Locating the Limits of Subject Representation in Libraries* (2002), to push against the "fundamental presuppositions" on which our information practices rest. (In Olson's case, she was focusing on subject representation within library systems.)

In the biodiversity realm, however, there certainly is an extent to which the "real" world plays a fundamental role in how and why we classify things the way we do. Biodiversity classifications are unique in that they engage with ostensibly natural-occurring objects that can be empirically examined and assessed for subsequent coordination in classifications. Biological objects can be assessed in many ways, using any number of traits: morphological, genetic, ecological, and so on. However, even though the act of classification is empirically grounded, there is no still no presupposed, natural order that can be naturally translated into representational classifications. Biological classifications are *arguments* in support of a certain arrangement of classes. Classifications are models, not mirrors. Each scientist will have a different take on what natural taxa exist and how these taxa are related to one another. This distinction between documentary and biological systems is a key one to keep in mind. The contrast between these disciplines, and how they verify their classificatory arrangement, can tell us a great deal about the subjective and representational qualities of representational systems more broadly speaking.

So, make no mistake, this book may take biodiversity as its main topic, but in service of exposing the mechanisms within classifications that make them such powerful cultural constructs. The problematics discussed in biodiversity classifications—and the possible ways forward I articulate—are applicable to any and all classification systems. Every classification exerts power, and so the lessons learned in this space are applicable to all similar representational environments.

Finally, one danger of writing a book about biodiversity science as it intersects with technology is that, after a short time, the cases represented in the book will become outdated. With that in mind, I have done my best to frame this narrative in terms that are, hopefully, more conceptual and less temporally located. There is certainly a danger that this approach may present the process of science in too condensed a fashion, or it may appear that I do not give full and proper justice to the complexity and nuances of the work of biodiversity scientists. It is important to note that what many biodiversity taxonomists take as obvious—that taxonomies are constructed and artificial, for example—may not be as clear to practicing IS professionals, or to the public at large. My approach here is to straddle the line between professionals and broad audiences that may not be aware of the disciplinary approaches of taxonomic work. Similarly, my delivery of the theories and literature of information studies may be equally reductive and, thus, may gloss over some of the divergent opinions in the field, especially in relation to issues of control and power broadly conceived. Throughout, I hope my respect for this biodiversity work is readily apparent. I am in no way presenting a case that intends to downplay the importance of biodiversity classification work or to claim that it is messily arbitrary. It is a science that produces taxonomies that are testable hypotheses. But even scientific practices are human at their core. The sciences are a series of epistemic cultures (Cetina 1999) that are subject to rupture and change over time (Kuhn 1996). It is the influence of cultural norms on the practice of science that interests me, much the way culture also influences the practices in IS. My goal is to show that, in fact, interpretative acts are exposed in the act of delineating and circumscribing species, and that within these moments of exposure we can better understand how concepts change, not only within the domain of science, but also within other areas of import, including classifications as they exist in IS.

CHAPTER TRAJECTORY

The chapters roughly follow the trajectory outlined in figure 0.1. Building on Patrick Wilson's notions of descriptive and exploitative power in *Two Kinds of Power* (1968), I invoke a deconstructing analytic that imagines the space of classifications to be (roughly) constituted by modes of power at the following levels (see figure 0.1):

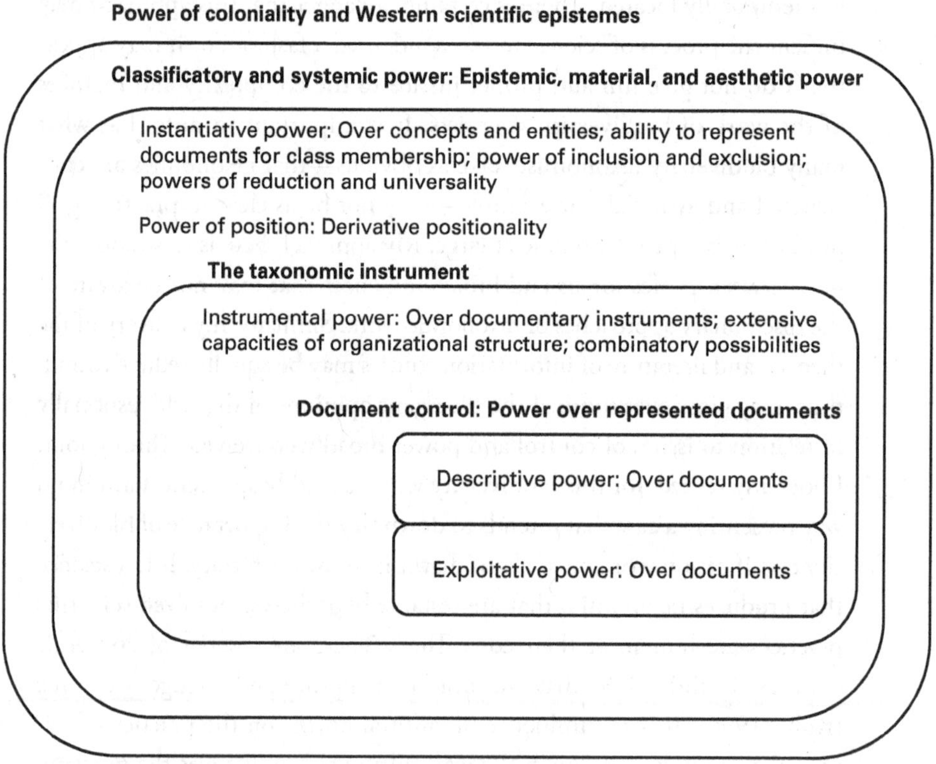

Power of coloniality and Western scientific epistemes

Classificatory and systemic power: Epistemic, material, and aesthetic power

Instantiative power: Over concepts and entities; ability to represent documents for class membership; power of inclusion and exclusion; powers of reduction and universality

Power of position: Derivative positionality

The taxonomic instrument

Instrumental power: Over documentary instruments; extensive capacities of organizational structure; combinatory possibilities

Document control: Power over represented documents

Descriptive power: Over documents

Exploitative power: Over documents

Figure 0.1
An analytic of classificatory power.

1. Power of coloniality and Western scientific epistemes
2. Classificatory and systemic power: epistemic, material, and aesthetic powers.
3. Taxonomic power (1): instantiative power: over concepts and entities; ability to represent documents for class membership; power of inclusion and exclusion; power of reduction and universality
4. Taxonomic power (2): Power of position and derivative positionality
5. Instrumental power: taxonomic instruments; internal and external extensive capacities.
6. Document control: descriptive and exploitative power as expressed by Patrick Wilson (1968).

Chapter 1 discusses the Catalogue of Life as the primary case of this text and shows how the "epistemic space" of these classifications are fraught with inconsistencies. In a composite environment, many different contributed taxonomies are juxtaposed in ways that defy traditional taxonomic norms. This creates vast gulfs in the practice of biodiversity between those that support such access-oriented measures and those that see these measures as obfuscating the function and practices related to traditional, internally consistent taxonomic systems.

In chapter 2, I begin by laying the theoretical groundwork for power within information studies as it relates to classificatory and representational spaces. A short narrative about the Australian dingo describes how classifications provide us the *power over* certain entities by virtue of classification, and the *power to* enact some kind of change—to the negative or the positive. I describe both the active and dispositional capacities of classifications. Active powers are those powers that are purposive, whereas dispositional powers are powers that are either unexpected or dormant until some user invokes them to some social end. I describe power as being both an individual power and a structural power that has an embedded, systemic quality that makes power difficult to identify in classifications. Classifications, as such, also make it seem as though the nature/human divide is real, but I argue that is not the case. I then describe how classifications have power over us in both material and epistemic ways. Classifications are epistemic in that they help us position ourselves in relation to other entities in the world, and as such, they are integral in helping us shape our lived and imaginary identities. Such identities depend on our relative positions in systems, which I call a *derivative positionality*, invoking feminist and indigenous theories of positionality to do so. I define the boundaries of classification in terms of their epistemic and spatial attributes, variously using the work of Miranda Fricker, Michel Foucault, Jane Bennett, and, especially, Henri Lefebvre to make these assertions.

In chapter 3, I briefly introduce the practical aspects of global biodiversity control by emphasizing and describing the processes by which we aggregate local data into global spaces and how this has prompted many epistemic challenges in the biodiversity world, connecting these issues to the Catalogue of Life. I end by asserting that social-ecological system approaches to understanding classifications, as postulated by Elinor Ostrom (2009), might be one analytic by which we can better understand the materialization of power in systems and may provide an avenue by which we can deconstruct systems to this end. In chapter 4, I turn to the power of instantiation, which is a core notion within the classificatory domain. I briefly describe how taxon concepts are formulated in biodiversity work and connect this process to the notion of instantiation theory within information studies. I then lay out the operational mechanisms that the Catalogue of Life, and others, has created to maintain control of taxon concepts-as-nomenclature over time.

Chapter 5 takes the structural "epistemic space" of classifications as its main subject and outlines how we see classifications as being constructed by way of a series of ontological and epistemological commitments about the natural world—and knowledge, more generally conceived. I first describe how the concept of forming species taxa involves artificially carving out categories (taxon, species, and so on) from an organic, continuous whole. I invoke scholars such as Alfred Whitehead, Kriti Sharma, and John Drupré to make the case that process studies has something to add here, focused as it is on understanding a system-oriented notion of being. My attention then shifts to the internal space of biodiversity classifications and briefly outlines how one's methodological and theoretical approach to taxonomy (by way of evolutionary taxonomy, cladism, and pheneticism, for example) produce fundamentally conflicting structures that are epistemically irreconcilable. I then end the chapter with a critique on *reduction* and *universality*, which are both necessarily core tenets of bibliographical, documentary, and biodiversity classifications.

Chapters 6 and 7 illustrate findings from my fieldwork with the Catalogue of Life and GBIF. Chapter 6 outlines the data-driven knowledge potentials of composite classification. By extending Wilson's two powers—descriptive power and exploitative power—I articulate *extensive power* as an integral attribute of these systems. Extensive powers work internally and externally. Internal extensive capabilities allow users to better understand the contours of the biodiversity data environment as it is expressed in the Catalogue of Life, or within any singular database space. With a global view of data, we can better understand where there are gaps in knowledge, as well as how we might bridge some of these gaps using data aggregation techniques. External extensive capacities include how the Catalogue is used in other systems and, by such a mechanism, influence the epistemic space far beyond its own boundaries. Finally, I discuss the power of prediction—with new aggregations of global data, one can use said data to make future-oriented inferences that would otherwise be impossible with siloed databases. Chapter 7 focuses on the contentions with composite systems, which are not minor or few. We see the impact of commingling different epistemic realities within one system. Problems arise with data control, with obfuscating local forms of knowledge, and with essential problems assessing the viability of data removed from its source. The chapter ends acknowledging the syntactic limitations of almost all traditional systems, including the Catalogue. Taxonomic methods based on genetic material, for example, do not always apply nomenclature to taxa, meaning that they are irreconcilable with the Catalogue's name-based formulation. This creates parallel, but incommunicative, streams of knowledge.

In the final chapter, I extend the conversation started in chapter 7 and move into the realm of the epistemic limitations of the Catalogue, and of all classifications that emerge out of a Western scientific tradition. I illustrate how the Western world has colonized indigenous knowledge via mechanisms that were (and are still) intended to "broaden our knowledge horizons." The result, however, is not the translation or integration of knowledge, but rather the essential reformulation of knowledge that is tantamount to continuing colonial epistemic violence. The chapter posits pluriversality as a worthy goal for Information Studies, particularly in designing the epistemic spaces of classifications. Though pluriversality has been proposed as a solution in other literatures, I offer a possible analytic and method to imagine new systems based in the field of design studies. I explain a *transition design framework* that can help us break free of our Western classificatory, univocal traditions. I end the chapter on the notion that classification justice is necessarily environmental and ecological justice, and if humanity is to change our epistemic frameworks, it is essential that information specialists play a part in refabricating our notions of what it means to "classify" at all.

CONCLUSION

In the end, no biodiversity taxonomic platform can serve all needs for all constituents; the question becomes how global control can be balanced with the flexibility required to do biodiversity work at the local level. Such flexibility can then be used to imagine and facilitate systems based on new epistemic modes of organization, radical connectivity, and collective dependencies. One goal is to think about how the historical, disciplinary, and theoretical specificity of biodiversity classifications can inform our own work and theories in information studies. This book, I hope, begins to show how the theories of information studies are applicable in realms far beyond our typical systems of concern. I also hope that my expansion of the concept of power helps us better intervene into classification systems such that we can make them more just and viable for multiple epistemic constituencies. Such an approach can, I believe, help us appreciate how the unique worldviews of myriad micro-cultures add positively to the inexhaustible (and beautiful) representational possibilities our information systems have to offer.

1 CLASSIFICATION SPACE REDEFINED

INTRODUCTION

This chapter introduces the primary case of the Catalogue of Life, a relatively new kind of classification structure that pushes against the traditional assumptions and functions of biodiversity taxonomies. Historically and traditionally, one of the hallmark qualities of classifications that organize organisms is that they are internally consistent, meaning that those who build them employ the same methods and organizing principles throughout the classification system. In this way, taxonomies are epistemically charged—they convey particular models of the natural world that serve as scientific hypotheses. Each taxonomic opinion is based on observed evidence analyzed in light of a unified and consistent set of classificatory commitments. More often than not, because the nature of taxonomic work is such that scientists will specialize in particular organism groups, taxonomies are usually limited to a single genus or other small subset of the taxonomic tree. For example, a scientist may specialize in beetles, or annelids, or crustacea, and the taxonomies produced in each of these areas mirror the taxonomic commitments of the scientists who build them. And because of the niche-oriented production of these taxonomies, many classification systems circulate in parallel with one another within the scientific community. As we will find, that there are many approaches to building taxonomies creates an environment where many different structures are produced that can be incompatible with one another. Despite this, these different classifications can be equally valid because they were built by taxonomists who use them to express a consistent, empirically produced hypothesis about how to order the world.

In direct tension with this traditional approach is the Catalogue of Life, which is focused on the efficient communication of species information with a global purview—that is to say, classification used primarily for data collection, communication, and retrieval. Specifically, the Catalogue is a particular kind of *composite classification*, which can be loosely defined as a structure that reconciles many local ("traditional") taxonomies within one space to serve as a unified taxonomic backbone structure for the organization of biological data. Backbone taxonomies, from this view, are user-oriented structures meant to serve primarily as ready-made hierarchical standards that can move biological data beyond what I call the "local" level of organization. In practice, composite classifications are also called consensus, common, or management classifications, terms I use interchangeably.

Not everyone is keen on these new composite structures, as many taxonomists feel the descriptive and hypothesizing paradigm of local taxonomies conflict with the epistemological commitments of consensus-based classifications. As taxonomies are brought together within the space of management taxonomies, various normalizing procedures are implemented to integrate these potentially conflicting subtaxonomic structures. Each taxonomy, ingested into the Catalogue as a database, must be negotiated with every other, meaning that conflicts naturally arise, and some taxonomies are altered to fit the management structure. To reconcile these conflicts, decisions must be made about which system, of two or more competing systems, is the best choice for inclusion into a common space. For example, if there is a conflict between two contributed taxonomies on the placement of the class Aves (the class level for birds), then the best taxonomy (as decided by the Catalogue editor) is based on prevailing scientific opinion. This exposes a significant difference between traditional and composite spaces: in the former, what is important is the correct placement of a taxon based on the internal requirements of the entire system to ensure that it is consistent; in the latter, the best taxonomic structure is that which represents community consensus within subfragments of the taxonomy, meaning, as a whole system, management taxonomies are likely *inconsistent* throughout. To focus my analysis, the bulk of this chapter describes the Catalogue's consensus structure and assesses how a "global view" of taxonomic opinion is manufactured. As we will see, taxon concepts are layered historical entities, so organizing these concepts within composite systems of classification adds to this complexity enormously. The underlying assumption of these common systems is that efficient data retrieval is preferred over fragmented and distributed data practices, which hinder data interoperability. Such fragmentation, some say, limits scientific breakthroughs and more emergent types of knowledge that can occur when research groups engage in broadscale collaboration.

Taxonomic aggregation of this nature presents a number of problems, including that the systems produced through this activity are less outwardly legible to anyone but the classification builder. If, as Tony Rees indicates, a classification is intended "to provide a recognizable navigation structure so that persons entering the classification can (hopefully) find their way to their particular organisms of interest" (2009), then how is a user to understand the Catalogue's internal intellectual structure? As Patrick Wilson (1968, chap. 4) makes quite clear, a good bibliographical instrument is one that people can understand, in order to appropriately navigate. Users should be able to anticipate (or at least be able to discover) the epistemological commitments of any given system. Consensus taxonomies, however, consist of many taxonomic databases, and thus this epistemological context can be far more difficult to understand, especially if a user is not a specialist in biodiversity work. This problem is especially important for biodiversity students or scientists who might wish to use the Catalogue for research purposes. Complete and sound taxonomic metadata from contributing taxonomies is most essential in this aggregated environment, as it allows a user to track back to a taxonomic source for authority assessment and construction principles. But metadata at this level of perfection is rarely, if ever, present (a fact that applies to all data sets, not just those of the biodiversity persuasion).

To fully understand the influence—and begin to explicate the conflicts—that these taxonomies present, in this chapter I also introduce how consensus structures act as social epistemological instruments, built as they are through intellectual compromise. "Social epistemology" as articulated by Margaret Egan and Jesse Shera (1952), Jesse Shera (1965), and Steve Fuller (2007, 2009) can help us understand the cultural function of these aggregative mechanisms. In a general sense, social epistemology examines how knowledge is shaped by social relationships and institutions (Goldman and Blanchard 2016). Thus, I am not only interested in how common classifications influence how people think about biodiversity in general and the problems introduced by common classifications to consensus proper, but also in what we should expect of classifications *at all*, given that they operationally define the state of knowledge within a discipline. Should they strive to represent divergence or manufacture accordance? Inasmuch as backbone taxonomies represent global systems, they represent a *kind* of consensus between scientific professionals. But what kind of consensus do they represent, and how are such mechanisms situated to tell us something that local taxonomies cannot? Certainly, consensus structures do not represent complete agreement.

At the same time, consensus structures align with how those in the field of information studies typically approach the collection and distribution of situationally relevant information to many different communities (Wilson 1973). This is to say that information from many domains can and should be accessible in a unified space to users and their varieties of contexts. Another assumption is that these composite structures are meant to be *general* systems, and that, perhaps, the audience for them is not, necessarily or primarily, the "local" communities that produced them, but rather, other allied fields that also require access to biodiversity information collocated at global and aggregative levels. So much of the discourse in knowledge organization and classification studies at the present is focused on the pluralization of the classification record (Andersen and Skouvig 2017). Universal systems are critiqued for good reason since, as with any system, they predominantly represent some communities and simultaneously ignore or harm others. Given these expressive qualities, classifications are unavoidably also political entities: they include and exclude, express and repress, facilitate and restrict. Some classifications are visible and some are layered behind a veneer of technics, algorithms, and system interfaces. The goal of IS, at least in part, is to make visible what is otherwise hidden from sight. Finally, universal classifications are fundamental to how we come to imagine, understand, and *position* ourselves embedded within a system of individuals, objects, laws, ideas, and ecologies in this world. Universal structures are about far more than mere schemas and access systems. If our goal is to collectively create classification and knowledge systems that are more flexible, diverse, and just—and I think these are and should be some of our main goals in practice—then pointed and fundamental discussions about what they do and do not offer are paramount.

However, to say that the Catalogue of Life, as a composite system, is a universal system is to stretch the term of universality beyond its reasonable limits, a consideration we will discuss in due course. First, an overview of how the Catalogue functions.

A COMPOSITE INSTRUMENT

> Our goal, therefore, is to provide a hierarchical classification for the CoL and its contributors that (a) is ranked to encompass ordinal-level taxa to facilitate a seamless import of contributing databases; (b) serves the needs of the diverse public-domain user community, most of whom are familiar with the Linnaean conceptual system of ordering taxon relationships; and (c) is likely to be more or less stable for the next five years.
>
> —MICHAEL RUGGIERO ET AL.
> "A Higher Level Classification of All Living Organisms"

One way to look at biological taxonomies is that they are meant, at least in part, to facilitate information retrieval. Even as far back as Linnaeus, systems of nomenclature and taxonomies were meant to improve information recall and facilitate memorization of species amid continuously growing stores of biological information (Müller-Wille and Charmantier 2012; Ereshefsky 2001, 366). As Patrick Wilson described in some detail (1968, chap. 7), these documentary systems are meant to be consulted—they must be an avenue by which a user can select the objects that meet a current need. This selection is predicated on the fact that, first, we have a bibliographical universe that is available in full, and second, that the documents therein are described in such a way to make them exploitable. This is to say, good advice about which document meets a certain circumstance or situation is dependent on being able to assess the full spectrum of organizational possibilities. Traditional taxonomies such as those described above, even if they fully satisfy the second criterion, fall short with regard to the first in that they usually contain only those taxa of interest in the particular domain in which they were produced. A taxonomy for weevils will only reference concepts or documents related to weevils and thus is really only of use to those who want information on this particular group. This is not a shortcoming of these descriptive instruments, of course, merely a product of how the production of knowledge is necessarily compartmentalized within the discipline of biodiversity. Given the fragmentary existence of these many descriptive taxonomies, a broader framework within which to control the full documentary universe is necessary. It is this requirement that the *consensus classification* is meant to satisfy.

A consensus classification is a comprehensive and pragmatic hierarchy intended to serve as a "framework for data integration" (Ruggiero et al. 2015a). These types of classifications—of which the Catalogue of Life is a prominent example but certainly not the only one—are intended to commingle the organizing structure of many traditional taxonomies in a singular space, with the intention of organizing data about, and potentially applying it to, numerous taxon groups. The Global Biodiversity Information Facility (GBIF) Backbone Taxonomy, for example—a management taxonomy of its own that builds on the Catalogue's structure—ostensibly brings data from anywhere together in one space for a multitude of purposes. To archive these distributed data, and to use them effectively as an integrated set of information, one needs a consensus classification as a practical tool to bring it all together. The taxonomic instrument as a whole is not intended to describe one unifying point of view but rather must reconcile many (perhaps, and often, contradictory) opinions in one space.

In a sense, consensus structures are a very familiar concept in IS—one can think of the Dewey Decimal System (DDC), for example, as a series of subclassifications brought together in one universal structure through which all produced documents can be organized and classed. The DDC is maintained with a coherent set of rules and guidelines that dictate how it can be revised over time. Each DDC class level—the main classes and subdivisions—is maintained as a separate domain-specific entity, each being updated and revised according to the documentation produced within the disciplinary area it represents. And because so-called general systems such as the DDC use academic disciplines as the primary method of classing knowledge (Langridge 1992, 21), we look to those disciplines (or perhaps more accurately, a discipline's produced documents) as the authority for that domain of knowledge: this is the notion of bibliographic warrant (Beghtol 1986). The recent revision of the Angiosperms in the DDC, for example, looked to phylogenetic and botanical literature for warrant to change their class structure (Green and Martin 2013). Biological consensus taxonomies function similarly: the authority for each taxon rests with the experts that research it, typically assessed by examining the relevant literature. In the sciences, however, a consensus classificatory approach is a much more problematic path, since its normative internal taxonomic structure can potentially conflict with the many other taxonomies that circulate in biodiversity spaces. Library classifications do not have the same epistemic and hermeneutic qualities as those that define biodiversity classifications.

A Constellatory Taxonomy

The Catalogue's management hierarchy uses the standard higher classifications of the traditional Linnaean system, including, but not exclusively, the ranks of kingdom, phylum, class, and order. Such an approach is suited to accommodating not only the largest amount of taxonomic information derived from contributing databases, but also to catering to the general public, who is most acquainted with this system (Ruggiero et al. 2015a). A consensus classification creates a backbone schematic that most easily facilitates communication between people and information systems. For example, if a person were looking for birds, they would look under Aves, and for beetles under Coleoptera. Having a unified standard for data communication makes good sense—it helps people navigate an otherwise complex system, because they are generally acquainted with its principles. For example, a few years back I moved to the Midwest from Southern California. In the Midwest, a popular supermarket chain is owned by the same corporation as the supermarket I frequented in Los Angeles. Luckily, both stores were organized more or less the same way, so I was able to locate fairly easily items I wished to purchase. The same logic goes for biological consensus systems: since most people know the Linnaean binomial nomenclature system, for example, it makes sense to keep that system in place to facilitate access. Versus, for example, creating a classification or nomenclature system that completely eschews Linnaean order.

As of the publication of Ruggiero et al. (2015a), 1.3 million of the then 1.8 million species contributed by subsidiary taxonomies were at the class level or below. Given this, the higher-level approach for the Catalogue's taxonomy was implemented to fill in blanks, so to speak, for taxonomies that didn't provide categories above the class rank, as well as to account for the intense disagreements regarding how these higher levels should be oriented if they were, in fact, provided. The end result is an incredibly detailed taxonomic hierarchy that provides a working and iterative structure on which more detailed databases can be appended (Ruggiero et al. 2015a, 10–55).

Like adding ornaments one by one to a tree, the Catalogue of Life attaches contributed taxonomic data to the consensus classification at the highest rank represented in the database until the tree is filled out and complete (figure 1.1). If a contributed Global Species Database (GSD) taxonomy (a taxonomy that includes all extant species for a group throughout the world, versus one that focuses on only a limited geographic region), for example, covers the Droseraceae family (Culham and Yessen 2018)—a group consisting of carnivorous sundew species including the Venus flytrap—it will be appended to the Catalogue at that family level, while the management hierarchy would complete the higher levels not included in that contributed database. This allows the Droseraceae data to be interconnected with the broader taxonomic tree. From this connecting node

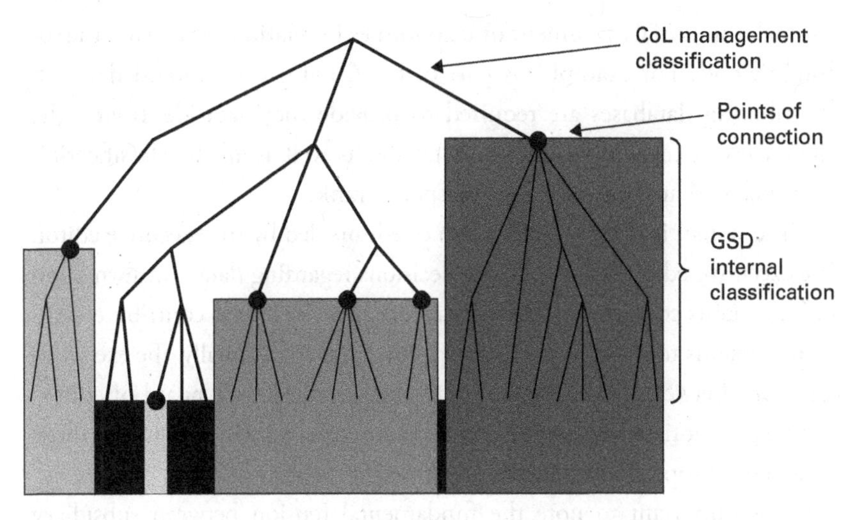

Figure 1.1
Schematic of the Catalogue of Life management hierarchy interacting with the GSD internal classifications. Original image label: "The Catalogue of Life retains the GSD's own classification below points of connection and uses the management classification above" (Species 2000 2016a). CC-BY 4.0, Catalog of Life, used by permission.

downward, the general composition of the Droseraceae of the contributed GSD is essentially maintained. Experts within the domain represented by a contributed taxonomy serve as the authority for its composition. It is sometimes the case that a database includes higher ranks that are not included in the overall Catalogue system, especially if the higher-level categories do not fall in line with the Catalogue's established hierarchy. In such cases, those higher levels are redacted before entry in the management classification.

The node of attachment for any taxonomic contribution is carefully and individually assessed through editorial review (Ruggiero et al. 2015a; Gordon 2009), and each species database is linked at only one node in the classification. Given the unique challenges posed by each taxonomy, no uniform standard exists by which this integration can take place. To reconcile the complexities and conflicts that arise from assembling these taxonomies, the Catalogue editorial team has articulated standards for certain required metadata elements to facilitate this reconciliation (Species 2000 2014). These standards provide a stable set of elements on which editors can make an initial assessment of taxonomies in relation to the larger taxonomic whole. For example, as part of the Catalogue's standard data set, contributing databases are required to provide metadata identifying the highest taxon covered within the database, as well as all the classification ranks below that point down to the species rank.

The system is managed by a team of editors, led by the executive editor. The executive editor handles all the decisions regarding data transformation and maintains contact with the various data providers that contribute taxonomic systems to the management classification. Additionally, the executive editor has between seven and ten individuals serving on a board of editors that helps negotiate with regional hubs and umbrella providers such as those located in Europe, China, and South America (Roskov 2016b).

It is important to note the fundamental tension between subsidiary taxonomic authority and editorial review. Not all taxonomists agree on any given taxonomic alignment. But assuaging these disagreements is less a concern for the Catalogue than is representing, as accurately as possible given the constraints, the diversity of taxonomic opinion within one consensus space.

On the Taxonomic Commitment to Access

The Catalogue's founder, Frank Bisby, believed that in order to truly understand the extent of biodiversity knowledge, a mechanism was needed to aggregate data in a unified and coherent manner. Then, and only then, would scientists be able to address gaps in species knowledge on a global level. The methodology of the Catalogue was motivated by the concept of access, not description, as the predominant taxonomic function. As the Catalogue's executive editor conveyed, "We were firmly dedicated to the position that we should provide users [with] a single view of taxonomy. A single, simplified, view unified through different codes: zoological, botanical, bacterial codes of nomenclature" (Roskov 2016a). The Catalogue was intended to "disjoin" taxonomic practices "from the desk" of individual scientists, facilitating a cross-taxonomic discussion in the unified space of a hierarchy (Schalk 2016a). Many users, after all, are less interested in taxonomic order specifically, and more interested in finding information about a certain species—particularly users outside the domain of biodiversity (Bisby 2000).

The process of creating a consensus hierarchy was difficult and complicated for the Catalogue; it lasted approximately two years, full of a great deal of debate and deliberation (Orrell 2016; Schalk 2016b). And this debate continues in many sectors of taxonomic practice. The editorial group knew that the purposes of the Catalogue would eventually be challenged as being antithetical to the function and commitments of description-based, traditional systems. Two primary concerns arose as central to conceptualizing the hierarchy: (1) how to best establish taxonomic authorities for each taxon group and (2) how to articulate the function of the classification in contradistinction to local, description-based structures. To address the first concern, each portion of the management hierarchy was assessed in consult with experts in the field, as well as through examination of more than two hundred sources of literature (Ruggiero et al. 2015b). Both the consulted experts and the literature acted as the warrant for a final taxonomic view that would ultimately be adopted over any other. Clearly identifying and citing taxonomic authorities was central to this process, primarily because, as Peter Schalk reminded me, no final product would match the views of all scientists, and thus an established hierarchy needed to be justified at every level—a process eventually documented in the article, "A Higher Level Classification of All Living Organisms" (Ruggiero et al. 2015a).

Regarding the second concern, the Catalogue is about *managing data* and maintaining this sharing functionality, not about managing internally consistent taxonomic opinion. It is a pragmatic structure. Although the Catalogue is generally "reflective of phylogeny" within its hierarchy (Ruggiero et al. 2015a, 2), it does not purport to be a cohesive phylogenic tree built to answer phylogenetic questions. For the Catalogue, constructing a taxonomy has nothing to do with a correspondence to some a priori notion of how evolutionary relationships should be presented—at least insofar as its classification is distinct from, or similar to, prevalent scientific opinion. Warrant is used to assess how each subtaxa should be presented, meaning that the Catalogue can potentially have many distinct opinions commingled in its classificatory space. As Catalogue executive editor Yury Roskov indicated (2017), it doesn't matter how contributed individual classifications are built but rather that they include the full, global reach of any given taxon and that it be up to date with regard to taxonomic method and opinion, citing appropriate literature for its construction. The assessment of any given contributed taxonomy often comes down, at least in large part, to the reputation of its contributor—both their individual reputation and that of their affiliated institution. In an environment where taxonomy cannot be pinned down and stabilized, all the Catalogue has to go on is whether or not a scientist has done due theoretical and methodological diligence in producing their classifications.

The underlying access functionalities of the Catalogue were articulated through careful examination of its potential use as a tool within the biodiversity world as well as in ancillary domains that equally needed access to the information. The Catalogue strives to manage a balance among four basic user groups: research scientists, policy makers, citizen-scientists, and educational users (Species 2000 2015a). Each of these user groups would come to understand the Catalogue in a vastly different way. Research scientists, for example, have used the database to understand the extent and distribution of species, as well as the valid names and synonyms for species concepts (Parr et al. 2012). The original impetus for the Catalogue is rooted in a policy-oriented mind-set, especially given that the instrument precipitated from the Convention on Biological Diversity (2003). In specific, the Global Taxonomy Initiative (Convention on Biological Diversity 2017a) identified how international scientific activities depend on taxonomic structures, including research on food security, conservation, and vector control. Concerning education, we see the Catalogue's backbone as integral to the structure of popular sites such as the Encyclopedia of Life (2018), which focuses on providing educational resources for a variety of curriculum levels.

One common thread among all these user groups is that their work, to greater or lesser extent, depends on the collocation of *all* species in one coherent space. With such pooling of intellectual knowledge, the Catalogue can facilitate global browsing and discovery mechanisms that any one local taxonomy would be unable to accommodate. Some of the possible access-oriented functionalities include providing (Species 2000 2015a):

- a single entry point into all distributed GSDs that make up its data space;
- a sense of global species coverage;
- universal searching tools for all extant species documentation and concepts, both inter- and intra-taxon;
- a hierarchy that can be fully browsed in tree form;
- a multilingual browsing mechanism accessible in eleven languages other than English.

Though these functionalities are productive, they point to the utility of the Catalogue as, primarily, a closed and self-referential entity—that is, these are all functions performed *within* the space of the Catalogue. Bringing the Catalogue into conversation with other, traditional taxonomies problematizes the veracity of its internal space. A host of other access functions require taking the Catalogue's taxonomic structure en bloc and using its backbone as the information architecture for other database and interface systems, such as the Global Biodiversity Information Facility (GBIF) and the Encyclopedia of Life (EoL). These systems incorporate the Catalogue's management structure into their systems, rather than creating a management taxonomy wholesale on their own.

Quite apart from the question of whether systems like the Catalogue or GBIF can be thought of as reflecting or conflicting with the aims of local (or description-based) taxonomies, one cannot ignore, as Slota and Bowker (2015) have conveyed, the broadscale impact these backbone structures can have on, for example, the articulation of conservation policy and new directions for scientific endeavors. In this way, systems like these directly shape how we come to understand the purchase and influence of this biodiversity knowledge on society. For better or for worse, global systems like the Catalogue provide us a mechanism for understanding the global state of affairs of biodiversity knowledge.

As a product of scientific collocation, the Catalogue is an instrument of *social epistemology*, as both a system that expresses a certain plurally driven, epistemic point of view, and one that modified how we understand the import of these data outside the biodiversity domain.

CONSENSUS TAXONOMIES AND SOCIAL EPISTEMOLOGY

The use of the term *social epistemology* deserves a bit of attention here, given its specific context within the history and literature of information studies. Social epistemology was introduced as a term within IS (Furner 2004; Fuller 2007, 69) through the publication of Margaret Egan and Jesse H. Shera's (1952) influential article, "Foundations of a Theory of Bibliography," which was later expanded on in Shera's monograph, *Sociological Foundation of Librarianship* (1970). In the former, Egan and Shera relate how the practice of bibliography had become fragmented as a discipline and thus needed more integrative approaches to communicate ideas both within and without the subdomain. The then-current field of bibliography (as it existed in 1952) was functioning as a series of separate groups, each "running around and around its own little circuit" (1952, 125). To illustrate a possible remedy, they introduce the metaphor of a railroad system to represent the pinnacle of information communication: an interconnected system primed to take advantage of the emergent knowledge that comes from cross-group information exchange. To Egan and Shera, complex forms of knowledge emerge from circumstances that comprise multilayered communication channels that exist *within* each specialist group, *among* these groups, and *between* specialists and the broader global information community (1952, 126).

In such a model, each specialist group could have a specific, and perhaps even conflicting, purpose from any other—and this diversity produces unexpected and sophisticated discoveries about the state of bibliographical knowledge. Part of this new approach was to take a more nuanced view of the *process* of information communication and to think about how the act of acquiring information had particular social impacts related to how one person (or group) could situate themselves "in relation to the total [intellectual] environment" (Egan and Shera 1952, 132). That the Catalogue embeds conflicting taxonomic opinion within its structure, for example, may not be a problem if we also understand that the biodiversity field of knowledge is equally defined by conflict of opinion. This metaphorical railroad system represented the total information environment—the entire documentary universe—and without the proper exchange of these "intellectual products" (Furner 2004, 793) within a bibliographic system, there could be no way to situate individual needs within a broader social environment. A critical impetus for the advancement of a discipline comes from the incisive insights that come from extra-disciplinary spaces. On top of this, we shape our research trajectories based on the findings of others and often act in response to their ideas. Thus, the more we share, the more we expand our disciplinary boundaries. Egan and Shera called this new proposed disciplinary emphasis *social epistemology*, which focused on the recursive "impact of knowledge on society" (Shera 1965, 107) and scientists.

A second significant proponent of an IS-centric notion of social epistemology is Steve Fuller, a professor of social epistemology at the University of Warwick, whose work emphasizes the specific material and documentary concerns of the discipline. Fuller defines social epistemology as the normative processes by which we "seek to provide guidance on how and what we should know on the basis of how and what we actually know" (2007, 177). He claimed that the institutional locus of the practice of social epistemology lies within the field of information studies, given that it is engaged in activities that validate and recommend access to, and use of, reliable and relevant information to society at large (2007, 6). Librarians—and, by extension, modern information professionals broadly speaking—"regard documents as sites for studying the multiply embedded social relationships among producers, consumers and objects of knowledge" (2007, 69). This curation of documents—assessment of a document's aboutness, value, quality, and relationship to a specific user—in turn affects how society interprets, and puts into use, the products of cultural and intellectual thought. The focus on the "document" as a core aspect of what we might call a socially geared epistemic agenda in IS makes good sense, especially since it is these collectively produced and organized knowledge systems (classifications, in this context) that shape our aesthetic and material relationships to the external world. Classifications help us see and outline what we know, and by what terms. In this way, the information scientist is not only a curator of the past, but also must be able to anticipate a need and direction for future intellectual work (Shera 1965, 109).

Of course, Steve Fuller (2007, 69–73) has certainly pointed out the dangers of a classification system oriented to its "dominant users," rather than, say, focused on the internal "organizational mapping functions" of its hierarchy. Moving away from the latter detrimentally alters a classification's epistemic orientations to one in which "success" proper is not defined as truth, but rather by a systemic alignment with expected public concerns and needs. Such an approach produces a reactive and service-oriented normative ethic, rather than one that focuses on curation and validity.

This existent dichotomy can be seen regularly in the discourse of biodiversity science. The divide usually manifests in conflicts between those who see taxonomies as hypothesis-driven arguments, and thus disagree with the aims of consensus structures, and those that see taxonomic backbones as user-oriented systems that facilitate the work of biodiversity informaticians. The former believe that sound inference comes from taxonomies that are internally consistent and, thus, facilitate more effectual work. The latter believe that such consensus structures actually unearth disciplinary inconsistencies and that they further a more global understanding of biodiversity issues. Of course, the reality of this situation is that the supposed "ends" of both of these approaches is to facilitate sound biodiversity work—the conflict lies in how they believe scientific consensus and agreement can come to be realized and represented in information spaces. Are the systems that we design to organize knowledge meant to impose order artifically onto the process of disciplinary knowledge production that is, by definition, defined by disagreement? What these critiques tell us is that, in addition to their use outside the domain of biodiversity work, consensus systems also are used to examine the shape of the knowledge *within* the domain itself. In the next chapters we closely examine these possibilities and critiques, so as to better understand these respective positions and the potentialities and limitations of consensus systems.

INSTRUMENTS OF DIVERSE OPINION

One function of this chapter has been to show how consensus taxonomies serve a particular purpose in biodiversity work that differs markedly from that of their "descriptive," or epistemically unified counterparts. These purposes align with access-oriented commitments that promise broad taxonomic representations across the full suite of taxa. Whereas description-oriented taxonomies are situated for local use, often and primarily within the domain of biodiversity taxonomic work, consensus structures allow that local knowledge to gain traction in domains that extend beyond their disciplinary applications. Consensus structures allow the discipline of biodiversity taxonomy to assess its internal contours—identify data gaps, for example—as well as to influence the circulation and constitution of knowledge beyond these borders, such as in policy work, conservation, and education. The Catalogue acknowledges this broader influence of its work as a tool for societal communication (Ruggiero et al. 2015a) and biodiversity knowledge implementation.

Returning to Egan and Sher's railroad metaphor, we can with very little effort relate and extend this notion to consensus structures within the biodiversity world. As mechanisms designed primarily for the retrieval of information—insofar as they are formed to understand the global extent of knowledge-via-data—they relate individual, description-based taxonomies within the broader biodiversity knowledge landscape. Backbones facilitate communication, however imperfect those structures may be. By constructing a centralized repository, we can better understand, for example, how biological knowledge favors certain charismatic species; what parts of the globe require more attention and investment; and finally, where our knowledge structures have erred with regard to syntactic and semantic qualities.

But the collocation of taxonomies into a unified space is not without its controversy. Any taxonomic instrument functioning on consensus paradigms—biodiversity, documentary, or otherwise—should be sensitive to the fact that these structures have drastic ramifications in a vast array of social spaces, within and without its domain of production, and with expected and unintended consequences. Our organizational tools mobilize bodies in the natural and social spaces that we inhabit. They also dictate which objects and organisms are prioritized and visible over any other. As Matthew Hull (2012, 260) makes clear, documents "generate larger-scale forms of sociality . . . not only directly as instruments of control but also as vehicles of imagination."

In "The Modernity of Classification" (2011), Mai recommends that we should begin to understand and interpret classifications, not as accurate models of the world per se, but as models of the way the world appears to those that construct them. There are many ways to view the organization of the world, and so our classifications should embrace this diversity. Extending this notion, some posit that there is a need for consensus structures in biodiversity taxonomy precisely because such diversity of opinion exists within the discipline and, therefore, there is a need to represent this discord in functional and operational ways. So long as taxonomists maintain divergent opinions, consensus structures can help operationalize this collective activity in structures that bring attention to these differences.

If the Catalogue exposes anything, it is that integrating one (unique, consistent, and local) taxonomy into a space of many (diverse and global) is a delicate process that questions the very function that taxonomies serve as hermeneutic instruments within biodiversity work. This becomes especially important because no single person can control what and how these structures are implemented or repurposed once they are made available to the wider public.

One of the biggest impediments to the ready communication of taxonomic knowledge is, as articulated by Thomas Orrell, the adherence to one and only one way to understand taxonomic practice. "Why does dogma overtake certain areas of the discipline?" Orrell asked, "Why is it that traditions can't be seen as separate from functional classifications of information?" (2016). Far too often, knowledge producers cannot see past their own methods and implementations to understand the utility of common systems that introduce a host of benefits that traditional, description-oriented systems simply cannot offer. The presence of multiple rules of nomenclature is an example of this dogmatic division Orrell mentions. Having only one rule of nomenclature would be much more effective, yet we have many. Our traditions maintain divisions and inadvertently also make it difficult to communicate the fruits of our research. What is it that stops us from agreeing on one classification, Orrell inquired, even as that one classification is meant to function differently from what we have come to historically expect from these structures?

Unearthing what common consensus structures can and cannot do is essential. First, a primary goal of the Catalogue is to provide a unified classification, not a universal one. The Catalogue's intent is to bring knowledge together for access and rearticulation, not to create a structure that must maintain authority in all domains of taxonomy. Second, the Catalogue is intended to provide a structure that can communicate diverse opinion, not (necessarily) one that seeks to provide and argue a singular taxonomic perspective. These distinctions are important in a domain that depends on taxonomies as argument-making instruments at the core of its work. Taxonomic work can and should function on two very distinct professional levels: one that requires a taxonomy to facilitate data transfer *and* one that can (and must) be used to serve as the structure by which taxonomic arguments are made. How can we divorce the expectation that a taxonomy must be an opinion from the idea that a taxonomy can serve primarily as a facilitator of information? The answer, in part, lies in better understanding the possibilities and limitations of the latter type of structures.

My hope is that unpacking these distinctions brings a broader awareness to their utility in various social spaces, even in spaces outside biodiversity work, such as in information studies. Systems like the Catalogue relinquish a certain level of conceptual and informational control in deference to necessarily embracing the fact that knowledge is diverse. Let it be known that, below, I argue that the space of the Catalogue is not as diverse as it might otherwise seem—nor is it truly plural in any secure or structural sense. This is especially true when we think about the kinds of data the Catalogue is unable to absorb due to its epistemic assumptions. However much one might value composite systems, it is my suspicion that the library world should look to biodiversity work for models on how to best represent knowledge, especially given the pressure placed on these institutions to embrace more pluralistic and diverse approaches to the organization of knowledge. We can learn from other disciplines as much through their trials and errors as through their successes.

Next we examine some higher-level properties of taxonomies and describe how taxonomies are instruments of power that exert in both material and epistemic ways. We then return to the Catalogue and examine its benefits and drawbacks in greater detail.

2 POWER AND POSITION

ON DINGOES AND CLASSIFICATION

The image of the Australian dingo has become something of a cultural and scientific flashpoint of late. In the United States, the dingo is perhaps most popularly associated with the phrase, *The dingo ate my baby!*, referring to a tragic moment in Australian history when a nine-week-old baby was reportedly killed by a dingo in the Uluru region while on a family camping trip. The baby's mother, Alice Lynne "Lindy" Chamberlain-Creighton, reportedly uttered a similar phrase to her husband, upon noting her child missing while also seeing a dingo fleeing their tent. The child's body was never found. Chamberlain-Creighton was initially convicted of the child's murder but was later released from prison upon a reevaluation of the evidence. Since this incident, many sources in US popular media, including *Seinfeld*, *Saturday Night Live*, and *The Simpsons*, have made light of this tragic situation. A film, *A Cry in the Dark*, also dramatized the circumstances.

The seriousness of this event and the prevalence of this vignette in popular culture have cast a shadow on the dingo, painting it as a social menace to be feared. In Australia, however, the dingo has a dual identity: as both a menace to farmers and a cultural and ecological icon for many indigenous groups. The overriding question is, should dingoes be protected, or killed as a pest to society? Do the farmers prevail, or the indigenous cultures that value the spiritual aspects of the organism? The conflict between these positions comes down to a simple (but actually not so simple) matter of categorization: Is the dingo an ancestor of a wild, untamed dog, or is it, in fact, a feral domesticated dog that has been reintroduced into the wild (Ballard and Wilson 2019; Smith 2015)? The former designates it a bona fide species that merits saving, whereas the latter qualifies the animal as a menace.

The dingo has long been a target of the Australian government because of its risk to livestock, pets, and humans; they were declared pests in the 1885 Amendment to the Marsupials Destruction Act. Additionally, current proposed revisions to the Declared Animal Policy under section 10 (1)(b) of the Natural Resources Management Act 2004, "Wild dogs and dingoes," allows for, and in many cases mandates, the controlled eradication of "the pest" through the use of baits, shooting, and trapping (Office of Parliamentary Counsel 2020). Most famously, however, is the Dingo Fence, built from the 1880s onward to keep dingoes out of fertile farmland (Allen and Fleming 2004; Allen and Sparkes 2001). In 2019, Western Australia's minister for environment, Stephen Dawson, went so far as to threaten that he "will make an order that determines that the dingo is not fauna for the purposes of the Act" (Bamford 2018).

Opponents of this eradication campaign contest the law on multiple grounds. In an open letter to Tim Whetstone, then minister for primary industries and regional development for South Australia, numerous academics pushed against the region's proposed revisions to the Natural Resources Management Act 2004 (Cairns et al. 2020). They rebut the decision based on multiple grounds, including that dingoes serve as an important member of the ecosystem, in that the species keeps the native kangaroo and fox populations in check. Poison baiting, they argue, also endangers legitimate working dogs and other nontarget species, further enhancing detrimental effects on the environment. On top of this—and most pertinent to the categorization issue at hand—is that there is no definitive scientific consensus on whether or not dingoes are "wild dogs" and thus not worthy of national conservation. In many areas, including Victoria, Australia, dingoes are classified as threatened and therefore protected under the Wildlife Act 1975.

This classification of the dingo is a story of (at least two) conflicting hypotheses (Ballard and Wilson 2019; Smith 2015). The first assumes that the dingo, introduced into Australia between 3,500 and 12,000 years before the present is not a feral wild dog, as indicated in legislation (Smith

2015). In line with this argument, the ecological niche dingoes filled upon entering Australia made them a unique species. Multiple points of evidence point to this fact, including skull morphological features and behavioral and breeding patterns, as well as genetic evidence (Smith 2015; Ballard and Wilson 2019). For the government to overlook this evidence means its policies are essentially endangering one of Australia's truly wild species. As outlined by Bradley Smith "As soon as the dingo arrived [in Australia], any influence of artificial selection ceased; it has undergone more than 4000 years of natural selection. If modern populations of dingoes were essentially the same as modern domestic dogs, that would suggest a rather remarkable lack of effect for natural selection shaping the dingo since its arrival in Australia" (2015, 74–75). This leads Smith to conclude that dingoes should be classed not as a feral dog (*Canis familiaris dingo*), as presently designated, but as a unique species reflected by the nomenclature *Canis lupus dingo*.

On the flip side of the coin we find proponents of the "wild feral dog" hypothesis. Under this assumption, the dingo began as a wild ancestor that was tamed in Southeast Asia before entering the ecology of Australia, where it was subsequently domesticated by human communities. This Asian domestication is the primary factor differentiating the second hypothesis from the first. Over time, domesticated dingoes strayed from human communities and began to live more self-sufficient lives in the wild as feral cousins to their tamed counterparts (Ballard and Wilson 2019, 4).

This dingo narrative is still very much in development in Australia, with little to no indication that the argument will be settled any time soon. In the meantime, however, the fate of the dingo as a biological, social, cultural, and natural object remains an open question. For the moment, they continue to be killed at increasing rates, which is unsettling on a number of levels given these disagreements. Irrespective of the eventual outcome of this debate, however, this story presents us with a series of complexities that are expounded throughout the course of this book: biological classifications have aesthetic, biological, cultural, and social forms of power. Multiple issues are at stake here, both ethical and economic, including whether we will "allow" an organism to exist in the world (that we have the power to make such a decision is sobering enough); whether tribal communities have a right to maintain a sacred population of animals (as if we have the right to adjudicate religious belief); and, finally, whether the farming industry is sustainable in the heartland of Australia (do domestic animals have more essential rights to live than wild animals?).

In light of these complexities, in this chapter we define how power functions both epistemically and materially in relation to classification systems. We also begin the process of imagining how we can model the operationalization of this power over the social and natural worlds.

POWER IN GENERAL AND INFORMATION STUDIES

It is important to articulate straight away what we mean by power, especially given that the term is so widely used within the context of information studies (IS) and that the concept will frame our subsequent examination of biodiversity classifications. Power is a basic and essential attribute of classifications, in that power exists regardless of whether or not we intend for it to be present. That is, even if you wanted to, you could not create a biological classification system that didn't exert some level of control over the natural world it is intended to organize. Control is one of the core reasons we classify. As George Lakoff wrote, "without the ability to categorize, we could not function at all, either in the physical world or in our own social and intellectual lives" (2012, 30). To categorize is to iteratively gain a better understanding of the world—to navigate that world, and to manipulate that world to our benefit. Categorization also helps us situate ourselves in relation to the world around us. Yet, despite its use in the IS field—to positive means and ends—the definition of the term "power" remains provokingly unspecific.

Generally, when we speak of power socially, we understand the existence of a dynamic in which Person A might have some advantage, control over, or privilege in relation to Person B. Metaphorically, one way to characterize having power is to imagine an individual having some authority, potential, or agency that is "over" or "beyond" some other person's authority, potential, or agency. In the realm of knowledge, a person might have access or be privy to knowledge in a way that puts then at an epistemic advantage in relation to other people. Power can also extend beyond the individual, to groups and communities. Minoritized communities for example, can be said to lack power in many senses. That disadvantage may be of language, of civil structures working to favor white epistemic constituents, or perhaps of economic hardship. Power, according to Miranda Fricker, is the "capacity we have as social agents to influence how things go in the social world" (2009, 9). In the case of members of a minoritized community, perhaps they do not have access to libraries or other social institutions that provide access to civic information. Community members are therefore less able to participate in democratic processes and have less capacity for bettering their lived circumstance.

Power can also be understood structurally, as part of established social infrastructures, technologies, and systems. As posited by Gerhard Göhler, structure and agency often work alongside each other and are complementary. "Human agency produces structures which simultaneously serve as the conditions for [the] reproduction of human agency in a continuous process" (2009, 30). When power is structural, it is often diffuse and not easily locatable within a given part of a system. In the case of biodiversity classifications, for example, the dingo might be classified as a "feral dog," but we may not have entrée into the processes that made this classification so. Policies and decisions for large systems are often made by many people and may involve following systemically derived processes and standards. Answering the seemingly simple question, "How did the dingo get classified this way?" can be far more complicated than it might seem. This makes power difficult to control and leverage in systems. Part of the function of this discussion is to expose where forms of power reside in classifications and how we might mitigate or enhance their influences.

We can ask this question, for example: Where do we put pressure on a classification to rebalance the distribution of powers to make it more just in relation to the wild dingo population? We might look to the forms of evidence used to delineate the taxon concept of the dingo, or whether it was properly articulated as a species taxon in relation to other organisms. We might also interrogate how much credibility the classification builder gave to indigenous forms of knowledge that hold the dingo as sacred.

Similarly, we can look to the governmental processes that facilitated the decision to use *this* classification of the dingo over any other. To make a classification more just, following the terms stated above, intervention is often required at multiple locations within the classification. Such change might require changing the attitudes of individuals (such as the particular scientist responsible for building this classification), or it might require more systemic changes that go above and beyond an individual, which might involve many actors (such as a reevaluation of governmental policy). Likewise, if we look to the consensus classification put together by the Catalogue of Life, we see that power is manifest in multiple locations: through decisions made by the executive editor; through a host of standards, policy documents, and partnership agreements; through the subsidiary policies of contributed taxonomies; through government regulations or scientific norms that dictate particular taxonomic needs; and so on.

Structural power also grows stronger the longer classifications exist, as they become more embedded in the social structures of our daily lives. In the case of the Catalogue of Life, the more databases that implement its management taxonomy as core data architecture, the more power it has over the biodiversity scientific environment. Species names, for example, are often embedded into governmental law, which makes changes to taxonomic names especially cumbersome. Let us take the case of the unassuming, tiny delta smelt and the surrounding battle over water rights in California (Savage 2015). Caleb Scoville's (2019) article, "Hydraulic Society and a 'Stupid Little Fish': Toward a Historical Ontology of Endangerment," outlines the complex interrelationship between the delta smelt (*Hypomesus transpacificus*), the Endangered Species Act, and the Metropolitan Water District of Southern California. A 2007 ruling by the US District Court of Eastern California impeded the rerouting of water from the Sacramento–San Joaquin Delta to Southern California based on the Endangered Species Act protection of *Hypomesus transpacificus*. The fact that the endangered species *Hypomesus transpacificus* exists at all was thanks to the increasing focus on the delta as an environment hot spot, which prompted an in-depth examination of the fish. Canadian ichthyologist Donald McAllister was responsible for distinguishing the pond smelt (*Hypomesus olidus*), which included the species currently inhabiting the delta, from the new delta-specific species (*H. transpacificus*) based on minor morphological attributes. In legal terms, the species *H. transpacificus* holds particular protection, but any change to this taxonomic name in the future can have detrimental effects on the conservation of the species. In this way, classifications find stronger and stronger footing in embedded national and global systems over time.

Importantly, systemic power is not always explicitly active, but can be activated by its potential dispositional attributes. For example, Fricker (2009) notes how a traffic warden may have power over a certain driver by virtue of *potentially* fining them for illegal behavior. Legal regimes, then, control our actions by passively imposing a power over us. Michel Foucault's (1995) notion of obedience relies on this kind of power—the power of potential surveillance that is a primary mechanism of control and discipline. Or, for example, if we think about classifications specifically: a hypothetical classification that is good (one that is representationally accurate to world conditions and also just) has the potential to help us advance our position in society by providing knowledge sources that attend to our current needs. Even if we aren't using the classification at any given time, it has the *potential* to help us. This power can be viewed in a positive sense, as in, "I have the power to make myself more knowledgeable in the fields of *X*." Or, "I am in the social position that allows me to pursue a graduate degree in luxury field *Y*." Likewise, classifications have the potential to be detrimental to our being and health. For example, if I were a member of the LGBTQ+ community in the early 1960s, seeking out information about queerness would present me with the social reality that I was, in fact, experiencing a medical "disturbance" and participating in deviant and paraphilic behavior (Adler 2017; Drescher 2015).

Classically, in the domain of IS, we often discuss the power associated with, for example, the descriptions, aboutness statements, and classifications we apply to documents, books, or resources. A central text in this vein is Patrick Wilson's *Two Kinds of Power: An Essay on Bibliographical Control* (1968). In Wilson's view, the activities of organizing and describing documents are closely aligned with the concepts of power and control. Wilson emphasizes that power *over* documents influences one's epistemic standing—a conclusion that becomes much more apparent in his later works on private and public knowledge (Wilson 1977, 1983). How adequately we describe the aboutness or content of documents is proportional to the quality, quantity, and usefulness of documents retrieved by some imagined individual. In this way, power over texts is power over knowledge. Knowledge-as-power, then, is power over cultural production and the ability to transmit that cultural identity from person to person and from moment to moment. Similarly, our goal when organizing concepts into classifications is, ostensibly, to take a world of entities (documents, organisms, digital images, and the like), to represent and relate them in some system, and then to provide bounded mechanisms for their retrieval. At each step we distill a world of chaos and fluidity into manageable webs of documentary simplicity. The process of distilling the complexity of the biodiversity world is to chronicle delicate fundamentally processual ecologies of organisms and to make sense of how each entity and group within that ecology relates, in some way, to another.

Key in Wilson's text is that the power is not inherent in or born of the bibliographical system itself, but that the power is, in fact, a quality imposed or provided by the person that does the describing. That power is then transferred to a person who subsequently uses that particular retrieval instrument to access information. Power is quintessentially a social phenomenon that takes place between individuals that participate within the same social domain. As such, power is used here in two senses: the power *to do* and the power *over*. The relationship between the *power to do* something and the *power over* something is intricate and complex and can be traced directly to classificatory systems and the organizations that design and maintain them. As Göhler states, the traditional distinction in the political sciences can be described thus: "*Power over* means power over people, enforcement of one's own intentions over those of others, and is thus only conceivable in a social relation. *Power to*, on the other hand, is not related to other people. It is an ability to do or achieve something independent of others. It is not a social relation" (2009, 28; emphasis original). But as Göhler concedes, the categories of *power over* and *power to* are inextricably linked. In one sense we can think of *power to* (latent and potential in its capacity and thus dispositional) as a precursor to *power over*. That is, one should first have the capability to perform some action (the knowledge and professional position to classify, for example) before any external relational power can be exerted. But the chicken and egg argument applies here, for a person with *power over* capabilities is afforded the *power to* by virtue of some social privilege or position.

In Wilson's terms, power is articulated from the perspective of both the user of bibliographical systems and the classification builder. A person can have what Wilson calls "descriptive control," which, as you might gather, relies on a document's applied descriptions (in a catalogue, for example) to concatenate documents according to some metric—say an author's name, a subject term, or title. On the other hand, a person can have "exploitative control," which is a mechanism of collocation that prioritizes only those documents best suited to a person's situational need (Wilson 1968, chap. 2). Wilson's *Two Kinds of Power* also serves as a kind of rule book on how to design bibliographical systems that are more just and effective for individuals. Wilson describes the power to distribute information adequately and equitably as a property of both an individual and a group of individuals (as in his hypothetical Supreme Bibliographical Council in chapter 9) (1968). What stands in the middle of these two extremes—the builder and the user—are bibliographical technologies that, in Wilson's hypothetically perfect world, would directly transfer justness through the seamless elaboration of descriptive and exploitative powers. Yet, we (and Wilson, I think) know full well that such mediative technologies live only in our imaginations; bibliographical and classificatory systems are far more complex than a direct line between builder and user. As described by Montoya and Leazer, "Wilson was generally deflationary and skeptical of the role of KO [knowledge organization] and whether it could accomplish its objectives[;] he did not analyze how KO could be effective for some people or some forms of knowledge, and ineffective for others, nor did he analyze similarly the constitutive components of KO as they relate to the differential effects on the use of knowledge" (2019, 162).

That the bibliographical system serves as a conduit for the power imbalances means that more critical examinations of the construction and distribution of power of and by classifications are necessary. To this end, we need a theory of power distribution in classifications as well as a method to decipher where this power lies in waiting.

CLASSIFICATIONS: PURPOSIVE AND DERIVATIVE POWERS

Classifications have material and epistemic powers that are active and purposive (as in, intentional on the part of the producer of classifications), as well as derivative (as in, consequences that cannot be foreseen by the creators affect an individual or group's identity-shaping capabilities). Even the list, the most basic form of classification, is a powerful cultural technology that wields a great deal of external influence. The position of an entity on a list—biological or otherwise—can have drastic ramifications for it in the material world. As Liam Cole Young notes, "Serious ethical and philosophical stakes emerge that demand investigation, particularly regarding the role of lists in controlling populations and subjecting human beings to power" (2017, 67). Thomas Mullaney's *Coming to Terms with the Nation* displays these stakes in great detail in terms of the classification of ethnicity in China as part of the country's 1954 Ethnic Classification project. Recent arguments for changing the Library of Congress subject terms for "alien" and "illegal alien" to "noncitizens" and "unauthorized immigration" also express these powers (Aguilera 2016).

Take, also, the IUCN Red List as a prime example, which iteratively lists the most threatened species on the planet. The scale spans species of least concern to those that are critically endangered and extinct (IUCN 2019). The Red List is intended to be used as "a straightforward way to factor biodiversity needs into decision-making processes" (IUCN 2020). The Red List establishes a baseline for better understanding how nature is being affected by humanity's actions, and thus places a focus on those species most at risk of extinction. The Red List ranking is also used as an agenda-setting rubric around which scientific studies are focused and by which conservation resource allocations are established (Barr and Wilson 2018). The presence of an organism on the list, such as our polarizing dingo, if ever included, can affect the actions that a government or a policy maker can or cannot take in response to certain species. If a species should be on the list, but is not, the ecological ramifications can be severe and long-lasting. But a list is just the first step in the operationalization of power as it is expressed within the medium of an organizational schema.

Acknowledging that classifications are constructed and purposive with regard to their social impacts is now fairly common in the IS discipline (and certainly in the biodiversity taxonomic community as well). We can thank the likes of Hope Olson (2002), Jonathan Furner (2009a), Joseph Tennis (2012), Melissa Adler (2017), and Safiya Noble (2018), among many others, for critically unveiling the mechanisms that lead to bias and social harms through and by classifications. This said, however, there still seems to be a social disjunct between *knowing* that they are constructed and actually attending to and offsetting their biased ramifications in our daily life. Part of this difficulty lies in the fact that rules for a classification system don't make themselves readily apparent. Good classifications, as Bowker and Star (1999) explicated, are those that integrate seamlessly into our experiential and epistemic fabric. Good classifications make it seem like they just *are*, and that they somehow represent a natural order of things. One of the lasting contributions of Bowker and Star's analysis in *Sorting Things Out: Classification and Its Consequences* is the approach that they posit for revealing the mechanics of classifications that are otherwise obfuscated from our direct view. "Infrastructural inversion" (Bowker and Star 1999, chap. 1) is the process by which classifications are deconstructed by means of interrogating their component, material parts, as well as the social contexts that influenced their design, construction, implementation, and maintenance. Infrastructural inversion sees classifications as a complicated intermingling of standards and networks that are influenced by various political and knowledge regimes (1999, 34). A suite of themes that can facilitate this analysis are then identified by Bowker and Starr; they include examining a classification's ubiquity, materiality, and texture; examining the historical narratives that led to its construction; investigating the politics of standardization and universality; and examining how all of these aforementioned attributes converge in points of both conflict and harmony. In this way a narrative of classifications can emerge by examining its component parts and their interactions.

The same type of deconstruction needs to be applied in the case of biodiversity classifications, to examine how power is exerted through the mechanism of classification and its attendant social and political contexts. Foremost in this analysis is that builders of classifications do, in fact, actively design classifications with certain epistemic slants. And just as Emily Dickenson's "certain Slant of light . . . oppresses, like the Heft" (Dickinson and Johnson 1997, sec. 258), these design decisions contribute to the impact these systems have on the external world and how they instantiate representations that then radiate outward and *define* a certain class of objects or subjects. To construct a classification is to situate entities in the external world by way of specific procedures, assumptions, and epistemic commitments. Part of this analysis includes ways for better understanding the organizational contexts of biodiversity classifications, which, among other elements, can help us understand the bodies of influence that dictate the intention, approach, and distribution of a particular classificatory technology.

If one were to take a Foucauldian approach to these classificatory spaces, we would see them as a discursive mechanism that filters the uncategorized natural world and "permits the visibility of the animal or plant to pass over in its entirety into the discourse that receives it" (Foucault 2007, 135). And, given that nature is "brought into" the discourse of scientific and popular language, the result is that we maintain control over the natural world, discipline it within the confines of predetermined standards, and manage these standards by virtue of laws, policies, and other acts of governance. On purely scientific terms, biodiversity science is an endeavor of love and a passion to catalogue: to give a name to the unique bodies of nature and to satisfy the same innate curiosities that have driven scientists, biologists, and philosophers for centuries. This same sense of wonder for nature drove Alexander Von Humboldt to the far reaches of the rainforest to redefine what it meant to exist within the complex interrelated network of the natural world (Wulf 2015). On a practical level, classifications are constructed to manage our relationships to the natural and, in an ideal world, used to balance human interests with the needs of a larger ecosystem we are a part of.

On the other hand, outside of these intended outcomes (such as classifications that take a clear side on whether the dingo is or is not a unique species known as *Canis lupus dingo*), a number of affects that classifications trigger may be unforeseen by the builders of classifications. In this vein, classifications have what we might call *derivative and dispositional power properties*, which are properties of classifications that can be *potentially* exhibited under the correct conditions or in new social contexts. Of particular note, the derivative effects outline the possibilities for identity formation. In the structuration of Miranda Fricker's argument, power is a capacity and a *potential*: "Capacity persists through periods when it is not being realized, power exists even if it is not being realized in action" (Fricker 2009, 10). This derivative power of classifications is latent unless explicitly activated by means of some social activity or policy enactment. This is not to say that the constructors of classifications are free of ethical obligations to these derivative effects, only that these consequences of their design and capacity are not necessarily foreseen. A scientist fifty years ago may have, with good reason, believed dingoes to be a wild canine and not its own species, but now, the social stakes for the dingo are far greater than that scientist could ever have possibly imagined. As Safiya Noble (2018) describes in relation to race representation in search engines, just because Google did not initially expect for keywords such as "Jew" and "black girls" to present primarily anti-Semitic and racist content does not mean that they are not culpable for the ill effects the search engine algorithm has on society. In fact, the responsibility for the missteps of classifications lies mostly with those who build them. This fact makes it all the-more essential for IS to attend to the power dynamics implicit in the classifications we build and to find mechanisms to intervene and locate technical levers to ameliorate any issues.

The ethics of classificatory technologies is one of our greatest information concerns of late, as the prevalence and comprehensiveness of classificatory regimes are growing more influential daily. The line between the correct choice and the ethical one is constantly blurred in the forum of biodiversity work and beyond. A scientist knows full well that how the dingo is classified can determine the fate of the entire species. To classify biodiversity is to implicitly enter the realm of political action and activity

(Youatt 2015a). The delta smelt showed us how this might occur. And while an ethical scientist should (and would) never let politics and government dictate the interpretation of taxonomic evidence, the ethical implications of this evidence-based work should not be overlooked. We will return to this issue in chapter 8 when we speak of Jonathan Furner's (2018) push for a veritistic turn in IS. That is, we should see classification as a tool for the delivery of information that is not only relevant, but also just. And this ethic should influence the way we build these systems. It is incumbent on scientists to intervene into conservationist spaces and help influence governmental policy. The ongoing tensions between governments and conservationists highlighted by the dingo and smelt scenarios are common. The importance of taxonomic classification goes far beyond the classificatory work at hand, and such derivative powers need to be of greater interest in the field.

CLASSIFICATIONS: EPISTEMIC POWER

Two distinct powers of classification are most pertinent to my broader argument: epistemic power and material power. To have epistemic power is to have the capacity to control how others think about, express, and situate themselves in the world and in relation to others. The backbone of this argument is that classifications help us conceptualize our relationship to the social and natural world, aid in the constructions of our individual and collective identities, and also dictate how our identities are *positioned* in contradistinction to other people, animals, plants, and objects. To be categorized as LGBTQ+, or disabled, for example, is to also experience daily reminders of one's minoritized and disadvantaged position in society. To be classed in these categories is to be limited in terms of both our epistemic potential and our material potential. In the material world, being transgender might impact our ability to use public restrooms in some states, not to mention the physical violence that people experience when expressing these identities faithfully to the world. In this circumstance, transgender individuals suffer from what Miranda Fricker calls an identity prejudice. An identity prejudice is "a label for prejudices against people *qua* social type" (2009, 4–5), which ultimately affects a person's ability to maintain social powers more broadly speaking. Materially, classifications can inhibit us from certain infrastructural or organizational opportunities and dictate how we navigate the social world. If some states get their way, to be transgendered may mean you are unable to use the bathroom of your chosen gender, which a person should have the complete right to do. Such derivative effects are central to the powers of classifications.

When we speak about epistemic powers, we are generally speaking about the impact classifications have on *people* and, more specifically, how their knowledge of the world and themselves is shaped by classifications. Miranda Fricker articulates how power might relate, or contribute to, two primary modes of epistemic injustice: testimonial injustice and hermeneutical injustice. Testimonial injustice is a form of "injustice that a speaker suffers in receiving deflated credibility from the hearer owing to identity prejudice on the hearers part, as in the case where the police don't believe someone because he is black" (2009, 4). To be categorized as a certain kind of person means that you are limited in terms of your epistemic potential, both internally and externally. In short, your opinion and knowledge are undervalued, and the result is that you are less able to enact social change or influence the actions of others or for yourself. To be subject to hermeneutical injustices is to be unable to apprehend social position by lacking the tools of "social interpretation" (Fricker 2009). As noted by Robert Montoya and Gregory Leazer, "Hermeneutical injustice creates the circumstances for the emergence of both epistemic marginalization and powerlessness. It also outlines the recipe for the continued marginalization of a group that remains unchecked unless other epistemic interventions are enacted" (2020, 32). In Fricker's mind, social and identity powers are heightened when one's epistemic injustices are ameliorated through active modes of systemic intervention and change.

Epistemic effects have to do with one's knowledge and identity, whereas material effects happen in the realm of the social world by virtue of our relationship with subjects and objects. When we speak about the material effects of classifications, we are asserting that they affect how a person experiences, navigates, and relates with the external world. Jane Bennett, in *Vibrant Matter*, describes "thing-power," a peculiar mental and phase shift with regard to objects that moves them from the realm of "stationary, inert thing" to the realm of the affective (Bennett 2010, chap. 1). Things become agents in their own right as they interact with the multitude of objects and subjects around them. Classifications are unique simply because they have the capacity to name what otherwise had no name and to identify it as an object, subject to certain social affordances. In one moment, a group of giraffes might seem to be of one species, and the next, it might be revealed that there are, in fact, four different species (Bercovitch et al. 2017; Fennessy et al. 2016). In this way, classifications manage our expectations about what *can* be in the world and how those objects or subjects might change the way we experience and interpret our experience of the world.

CLASSIFICATIONS: SPACE

How does the space of a classification exert its impact on the space of the lived, social world? If we look back to the anecdote of the dingo that opened this chapter, we see how powerful classifications can be when thought of in practical, ecological terms. The placement or position of an organism in a taxonomy has direct implications for the negotiation of their occupied space in the external world. Our class determinations save or eradicate the lives of species, which, however we may feel about any species in particular, is never a power we should wield lightly. There is a direct, if unanticipated, connection between classification and our lived, physical space. In documentary or bibliographic classifications such as the Library of Congress Classification System or the Dewey Decimal System, classifications are intended to organize *all* knowledge into universal structures, which is itself a form of control and power. Organizational decisions such as what to exclude, how to label entities or concepts, and the overall intellectual architecture of the classification are all spatiotemporally contingent, based on what culturally makes sense at that time. We know, however, that once we start creating hierarchies in these systems, some concepts are obfuscated or excluded, while others are brought to the forefront and emphasized. Over time we see subjects come and go, depending on current social norms and ideologies, such as Tennis's examination of the subject term for eugenics

(Tennis 2012) or the inclusion of homosexuality as a pathology in the Diagnostic and Statistical Manual of Mental Disorders–2 (Drescher 2015).

Melissa Adler also expands on this material reality in her book *Cruising the Library: Perversities in the Organization of Knowledge*, where the categorization of "perverse" terms in the Library of Congress confines and constrains how we access and contextualize information on these subjects. In this sense Adler rightly believes that classifications are a technology of power that can potentially imprison domains of knowledge in institutional regimes such as the Library of Congress Classification (2017, 152). Adler's poetics of the body is enticing here, for in her examination, the body of literature and the constraints placed on it within classification technologies affect the way we interpret and experience our physical bodies. The line between the representational space of the classifications and the space of the experiential, "real," and social world is erased—a fact made uncomfortably visible when thinking about the inevitable fate that dingoes in Australia seem to be up against. Ronald Day proposes a "user" model that "views subjects and objects as co-emergences mediated through co-determining, contextual (or 'structural') affordances and through in-common zones of mutual affects" (2011, 86). In this way, context remains paramount in understanding how the "user" might benefit or be affected by being classed in particular ways.

Rebutting the false impression that classificatory technologies are truly representative of some reality is an ongoing task of the information professional—particularly those working in classifications. Tim Hawkinson's play on perspective in his sculpture *Shrink* is a useful place to investigate the connectivity between objects and their representations in classifications (Rinder et al. 2005, 152–153). The sculpture involves a wooden chair whereon microfilament strands are attached at regular intervals along the full outline of the chair (see figure 2.1). These microfilaments then meet at a "vanishing point" tethered in place by a stick emerging from the chair's frame. Tiny pieces of wood from the chair were then carried along the microfilaments to this vanishing point, resulting in a miniature replica of the chair seemingly suspended in midair. The piece is evocative precisely because of, as Lawrence Rinder notes, its "self-reflexive insularity,

Figure 2.1
Shrink, by Tim Hawksinon (Rinder et al. 2005). © Tim Hawkinson, courtesy Pace Gallery. Used by permission.

turning back on [itself] in an echoing loop of signification" (2005, 19–21). On one hand, you're struck by the beauty of the sculpture's visual symmetry and the ease at which the chair can be "modelled" using classical Renaissance-esque visual techniques of mathematical precision. Yet, on the other, while the model is precise, the interdependence between the representation and the object are tenuous and fragile. The connection between the sign and the signified is put into question as minor imperfections in the small model chair present themselves over time (I first saw this sculpture at the Los Angeles County Museum of Art in 2005, and by that time the wood fragments making up the suspended miniature chair had already begun to slip, highlighting the fraught relationship between reality and simulation).

In many cases, the relationship between the "represented world" and the "representing world" is assumed to have some degree of congruence, truth, or consistency (Rosch, Lloyd, and Social Science Research Council 1978, chap. 9)—that the space *there* is representative of the space *here*—but this cannot be further from the truth. We see this equivocation at work constantly, especially when we think about how the natural world is organized into species taxon compartments that fit, ever so neatly, into an evolutionary hierarchy. The world is, in fact, messy and continuous, and certainly not situated to easily compartmentalize in any "natural" or essential sense. And yet, the classificatory representations we produce display them as if they are. On the whole, this is not necessarily problematic—after all, scientists take great pains to make biological classifications accurate, base results on evidence, and go through peer review like any other discipline. And every biodiversity taxonomist understands that classifications are constructed—the problem lies in the fact that users do not always acknowledge the same. But cognitive dissonance at the point of use creates a circumstance in which, when classifications are not questioned for the power they wield, enormous social problems are sure to follow.

Not all control is nefarious at its root, of course, particularly when we discuss those classifications produced in the taxonomic sciences. Taxonomists build classifications according to particular theories and methods, irrespective of their use. As Foucault discussed at length, the control of space within society is structurally embedded in the institutions built to discipline bodies, knowledge, and their attendant activities. Whether through the guise of a prison or of a school, space is used to situate people, dictate their movements, and manage their learning. Jeremy Bentham's panoptic vision has now widened to include video surveillance, mobile tracking, and other locative media to further assess and direct the way we go about our daily lives. No doubt for good reason, surveillance in this sense evokes the negative connotations associated with *power over*. In *The Order of Things: An Archaeology of the Human Sciences*, life becomes the operationalization of what can be seen and nature becomes an *object* of knowledge that can then be transferred into language and inputted into a predetermined artificial system (Foucault 2007). This mechanism of control has to do with the establishment of order from chaos; to fix our gaze on singular objects that then become the seed for the study of natural history.

With all of this said, the compilation of lists such as the Catalogue of Life are scientific endeavors intended to both celebrate the diversity of nature *and* seek out the inherent natural relationships between species. But while the intentions are pure to the goal of advancing science, the act itself remains a means of control no less than in Foucault's world—though the outcomes of biodiversity work may be far more beneficial than the dangers presented by unprecedented regimes of surveillance. What biodiversity classification does do, however, is normalize nature into structural brackets that consist of standardization, temporal control, and mediated, carefully measured hierarchical arrangements. As Foucault states,

> In short, the art of punishing, in the regime of disciplinary power, is aimed neither at expiation, nor even precisely at repression. It brings five quite distinct operations into play: it refers individual actions to a whole that is at once a field of comparison, a space of differentiation and the principle of a rule to be followed. It differentiates individuals from one another, in terms of the following overall rule: that the rule be made to function as a minimal threshold, as an average to be respected or as an optimum towards which one must move. It measures in quantitative terms and hierarchizes in terms of value the abilities, the level, the "nature" of individuals. (Foucault 1995, 182–183).

What is important to adopt from Foucault's structure for our purposes is the nominational aspects of biodiversity work—the application of names, the delimitations of species taxa, the mediated and methodological approach of the work, and the production of relations through the application of particular metrics. Looking beyond Foucault, however, let us look to Henri Lefebvre and his thorough conceptualization of space to better understand how the representational space of the classification can be understood to relate to the lived experience of our natural world.

Henri Lefebvre is widely considered one of the major motivators for the "spatial turn" that transpired in the social sciences in the final decade of the twentieth century (Zieleniec 2018). One of Lefebvre's contributions to twentieth-century philosophy is his assertion that space is the essential and foundational element on which all means of production exist and interact. According to Lefebvre, Marx understood "things" as embodying the "intersection of social relations and the forms of those relations" (2011, 81). Lefebvre also believed that space was defined by the interrelation of social actions, subjects, and objects. In *The Production of Space* (2011), Lefebvre tackles what he sees as one of the great problems in the philosophical field at that moment: until Lefebvre's publication, philosophers had a fairly flat and underdeveloped concept of space as a theoretical field of inquiry.

According to Lefebvre, on the one hand, you have space in the Cartesian tradition, wherein space was an "object opposed to subject" (2011, 1)—an absolute space separate from the mind. As expounded by Chris Butler,

> Accordingly, [Cartesian space] could be reduced to a set of "coordinates, lines and planes," capable of quantitative measurement. This Cartesian account was supplemented and complicated by Kant's understanding of space and time as *a priori* categories that theoretically placed space within the realm of consciousness. These two primary influences have established a dominant philosophy of space that ontologically treats it as an empty vessel existing prior to the matter that fills it. (2014, 38)

Kant's notion of space is inherited by subsequent philosophers, juxtaposing a "mental thing" or "mental place" with the Cartesian notion of physical space (Lefebvre 2011, 3). But to Lefebvre, the definition of a "mental space" had not yet been clearly articulated, particularly as it related to social spaces. Because of this lack of clarity, the study of space became subsumed within studies of epistemology, where it grew stale and unattended, general and unarticulated. To Lefebvre, space within the mind was "fetishized and the mental realm comes to develop the social and physical" notions of space as well, flattening the nuances between the two domains (2011, 5). Because of this, the gulf grew between the mental and practical space (defined as both physical and social spaces), to the detriment of cultural ideologies. But, as noted by Christian Fuchs, Lefebvre believes that "space is neither subject nor object" (Fuchs 2019, 92, 135). Rather, space is a "social reality," and "a set of relations and forms" (2019, 116) and encompasses both the generative potential and practical limitations of cultural and social production. Space *subsumes* products and their interrelations (Lefebvre 2011, 73). Lefebvre's contribution, then, was to establish a model—"a science of space"—by which the realm of the physical and quantitative could be reconciled and brought into conversation with the mental and social realms of space. This move, then, can help us articulate one approach that shows how the representational space of classification relates to the space of the external, natural world.

Lefebvre's philosophy of space is premised on a spatial triad that consists of three distinct yet interrelated concepts: spatial practice, representations of space, and representational space (2011, 33). These three spaces roughly correspond to the three areas identified by Lefebvre as lacking unity: the physical (or perceived), the mental (or conceived), and the social (or lived) (Butler 2014, 41; Lefebvre 2011, 40). Spatial practice can be thought of as our daily patterns of physical movement. This is our perceived, "real" space and can be "evaluated empirically" (which is distinct from representational space, which is abstract and in the realm of memory and intelligence). Practice is the sum total of our mundane activities in space that maintain social life, bring individuals together, and form cohesive fabric of activity. A representation of space is "conceptualized space, the space of scientists, planners, urbanists . . . and social engineers . . . all of whom identify what is lived and what is perceived with what is conceived" (Lefebvre 2011, 33–34). It is the world of models, measurement, and precision that ostensibly act a mechanism to make sense of the many complex layers of the lived, embodied space. While physical space is infinite in its variety, representational space, as I understand Lefebre's perspective, must be delineated, planned, and digestible at a given moment. Classifications live in this space: they represent, nominate, and quantify the visible, and render our experience of nature into metaphoric, sensible apparatuses. Importantly, representations are tied to institutions (Butler 2014, 40), and given that Lefebvre highlights the sciences in this space, the focus is on the inherent power and politics in the artificial (yet careful) arrangements that define this space. Finally, there is representational space, or the space that transpires from processing the *lived* experience. This is where creativity, artistry, and "complex symbols are linked to hegemonic forms . . . and social resistance" (Butler 2014, 41). Representational space intellectualizes the lived and represented world as the space of memory, history, art, and belief. Representational space is where aesthetics is born and coded for communication. This space is dynamic and less ordered and linear than the physical spaces we encounter because it is not privy to the laws of the physical world (Lefebvre 2011, 39–44).

Each space in Lefebvre's triad works in tandem and recursively with one another. How we move in space has been dictated by the plans of architects and planners, and so too has our physical and represented space been integral to the way in which we make sense of our surrounding and coalesce our experience into meaningful narratives. How the sciences deconstruct and model our world and how we imagine our world related to these models dictates our personal motivations, our politics, and our individual and collective sense of identity. Directly connecting the space of practice and the space of representation is important for our purposes, particularly because it shows that classification is a material concern in our social lives: "Classificatory space is social, and perhaps more importantly, *produced* by way of the interactions of variety of social structures, policies, laws, and actions" (Lefebvre 2011; emphasis original).

BIFURCATION

It is important to note that, for the purposes of this narrative, the space of "society" and the space of "nature" are one and the same. As organisms ourselves—socialized as we are—we must shift our consciousness of humanity in ways that see us in concert with our natural surroundings, rather than apart from it. Classifications, however, at least partly because of the limitations of graphical spaces, will often support the unintended bifurcation of the two. Alfred Whitehead pushed against this bifurcation in his famous *The Concept of Nature* (1920). Although power is, indeed, enacted by an individual person, group of people, or system (which is, ultimately, created by people), the recipient of its affects can be human or nonhuman—and, in particular, can be the flora and fauna that subsist and depend on societal actions in the "natural" surround. The reality is that nature in its purest sense—that domain of the world that is not humanity-dominated as cities, towns, and the like—does not really exist "outside of human social impacts" (Burke and Fishel 2019, 87). The discourse within IS regarding the exertion of informational and classificatory power should not exclude the very natural environment within which we are embedded. There is room for organismic and environmental justice discourse in the discipline to better orient classifications as spaces of social and natural import. As Burke and Fishel emphasize, many elaborations on the concept of power—especially those in political theory, global politics, and international relations—fail to attend to these human/nonhuman relationships (2019). As Bruno Latour indicates, when we treat ecosystems and the natural world as something separate from the human we are essentially "proposing . . . that an arbitrary portion of the actors will be *stripped of all action* and that another portion . . . will be *endowed with souls* (or consciousness)" (2017, 49–58; emphasis original). The very act of bifurcating nature from society is an act of will to power that implicitly places the power in the hands of humanity, over and above the realm of nature. In this scheme, the human realm supersedes the natural category in terms of agency, importance, and priority.

In this light, if we think of Lefebvre's triad, the production of social space is, in fact, *also* the production of natural space. As Lefebvre states, "But today, nature is drawing away from us, to say the least. It is becoming impossible to escape the notion that nature is being murdered by 'anti-nature'—by abstraction, by signs and images, by discourse, as also by labour and its products" (Lefebvre 2011, 70–71). Even the very act of protecting nature through the establishment of national parks and reserves filters nature through the discourse of our social, political, and conservatory actions—and thus nature has no choice but to exist *within* our control rather than apart from it. On top of this, our spaces of classificatory representation allow us to minimize and compartmentalize nature, take hold of it, and carry it around in the guise of maps, guidebooks, and catalogues of species. As Susan Sontag said in relation to photography, "To photograph is to appropriate the thing photographed. It means putting oneself into a certain relation to the world that feels like knowledge—and, therefore, like power" (2001, 4). The mere act of representation is to relate to the represented on social terms and often within an imbalanced relationship with respect to power.

DERIVATIVE POSITIONALITY

It is not enough to establish that the spaces within and without classification are mutually constituted; it is also important to ask the question: What does it mean to stand in *this* space, or to be represented in *that* space? That is, how is one's positionality in the social world affected by their derivative position in represented space? One of the most consequential impacts of classification systems is the extent to which they *position* representations and codify them in particular contexts and at a given point in time. Positionality is a central concept in Wilson's *Two Kinds of Power* (1968). Wilson discusses the production of subject terms as a defining descriptive point required to identify situationally appropriate content within documents via bibliographical systems. For example, if we wish to find books on the subject of dingoes, we would be wise to search for books with a subject term "dingo," a term inclusive of both *Canis familiaris* and *Canis lupus familiaris*—both possible taxonomic names for the dingo depending on whether one believes it to be a specific species of dog or a subspecies of the wolf (Library of Congress 2020). For a truly comprehensive search, one may also need to seek out the "dog" heading, given that some cataloguers may not have applied the more specific "dingo" moniker. Examples such as these are everywhere, particularly because applying subject terms to documents—essentially establishing the *aboutness* of the text—is notoriously one of the most difficult ways of "assigning positions" within an "organizational scheme" (Wilson 1968). One must assess what the content of a document might be *about*, while also simultaneously assessing how a user might enter the catalogue and search for the document. Importantly, Wilson chose to title his chapter "Subjects and the *Sense* of Position," and I'd like to highlight in particular his use of the word "sense." To "sense" position is to interpret, to provide meaning where it might not otherwise reside, and evokes a notion of inwardness, sensing, and "feeling" as if something should be classed in *this* position or *that*. As discussed above in the context of Lefebvre's triad, if we think about the tight connection between lived space and representational space, we can see there is a connection between how we think about positions in the space of classifications and how these positions interplay with and affect positions in the space of society.

One's complex identity in the lived, social world is significantly affected by the intersectionality of their many representations in the classificatory, represented world. The argument here is not new, of course; we can look to a core text in IS such as *Indexing It All* (2014), by Ronald

Day, which, at least in part, unpacks the ways documentary technologies (library information systems, social computing intermediators, database indexes, algorithms, and the like) impact our social and lived identities by way of fragmented indexical systems. Information systems, sociotechnical apparatuses that they are, act as "constructing infrastructures for both producing subjects and objects" through the indexical relationship of users and objects, and their representations (2014, 41). These systems have become so pervasive in our social and political economies that they have become our primary social mediators, which, in turn, define both our individual identities (our subject-ness), and our relations to other people (subject-to-subject interactions). Further, these indexical realities play a far greater role than our embodied activities in social space. Returning to our opening example, any one dingo in the wild, for example, is affected by its classification (its representation) *far more* than it is affected by the sum total of its individual attributes. Our descriptions in classifications are spatiotemporally unrestricted, while our physical being is defined by the opposite. In some senses, these representations are also normative in scope, which has implications in terms of the magnitude of their influence. Normative because, if the dingo is classified in such-and-such a way, then the impact is not on one or a few, but on *all* the external natural entities that fall within that descriptive category of the species taxon concept. Through this radiant normativity we can better understand how derivative representations are also artifacts with materialities and affects in their own right. This is to say that representations-qua-classifications have *agency* within the world to a degree that is often not acknowledged in the course of our daily lives. They situate natural objects among each other, and inform our social, economic, and political relationships with particular species.

This is the way derivative effects should be understood within the space of classification systems, given that they, too, are documentary, indexical systems in their own right. Building on this, the goal is to further define how our classificatory identities are equally complex intersectional concepts, much in the way our lived, social identities are positioned at the intersection of many socially prescribed systems that coalesce to produce a seemingly coherent identity whole. I adopt the term *derivative positionality* here to highlight that representations within classifications are not only representational extensions of the physical ("real," external) objects and beings they classify. They are material and affectual structures in that they help form our epistemic understanding of our lived, social position. One's positionality (qua representation) in classifications supports and solidifies one's actual positionality in the social world. This kind of positionality is derivative in that it is primarily representational, but such representations nevertheless remain tethered to the identity, fate, and position of what is being represented. One's representations work recursively in that, as classifications change, our ability to negotiate the world and construct our identities are affected and also change. And, as our existence and identities evolve, so too do the positions of our representations in classifications. This is a prime example of Ian Hacking's "looping effect" (2007). If we create a new species, for example, that species is now a legitimate entity in the realm of policy, conservation, and the like. And, with a name, it can now subsist within a broader index of species taxa that bring to bear individuated consequences.

Of course, the distribution of power in social spaces is a relatively familiar and easy way to understand how our position in society is dependent on our individual, constructed social identity. Our privilege, mobility, and eventual success can be measured by the ease with which we can move from one social position to another, presumably more advantageous and beneficial, social position. In social spaces, variables such as education, social networks, inherited financial capital, ethnicity, and gender allow us to inhabit a specific position with respect to other individuals. This position, then, opens us up to, or restricts us from, a variety of social opportunities or benefits. Melissa Adler's (2017) argument illustrates how classifications discipline bodies of knowledge by positioning certain concepts in ways that degrade or bias a certain subject-identity (whether that subject term be paraphilia, sexual perversion, or homosexuality). A central power of these positioned representations is that they are more mobile, delicate, radiant, and accessible to the actions of others than are the individuals they represent. Derivative representations can "live" many places at once, and each is privy to different online or material contexts, which illustrates the strength of derivative forces as they function to construct the identity of what they represent.

Many may note that the term positionality is not often used in the context of classifications in IS literature. The term has a strong scholarly context in, for example, identity, feminist, queer, and indigenous studies, and is associated with standpoint theory epistemology. This more socially contextualized definition of positionality is equally as pertinent in this narrative, as it relates to how positions in classification also function in relation to our lived experience. The interpretation and methodological implementation of standpoint theory varies widely depending disciplinary context, but, as Elizabeth Anderson explains,

> Classically, standpoint theory claims that the standpoint of the subordinated is advantaged (1) in revealing fundamental social regularities; (2) in exposing social arrangements as contingent and susceptible to change through concerted action; and (3) in representing the social world in relation to universal human interests. By contrast, dominant group standpoints represent only surface social regularities in relation to dominant group interests, and misrepresent them as necessary, natural, or universally advantageous. (2020, Section 2)

In this vein, one's individual perspective, social position, and historical context not only influence but also define their epistemic stance and authority. It is a means of empowerment. To adopt standpoint theory is to position yourself in contradistinction to the material and external modes of power that subject us to distributive, representative, and other forms of social inequalities. One flavor of feminist standpoint exalts individual perspective in the face of "androcentric, economically disadvantaged, racist, Eurocentric, and heterosexist conceptual frameworks" that "ensured systematic ignorance and error about not only the lives of the oppressed, but also the lives of their oppressors and this about how nature and social relations in general worked" (Harding 2004, 5). In *Disciplining the Savages: Savaging the Disciplines* (2007), author Martin Nakata notes how the power and strength of an individual epistemic stance—one that may be generally viewed as underrepresented or disenfranchised—is constructed by way of many cultural and social elements that "interface" with lived experiences, historical memories, and individual historicities. Power dynamics between a colonial past and a Western future merge in Nakata's own identity as they do for other Torres Strait Islanders—the indigenous population in Queensland, Australia. Islanders "operate on a daily basis in a space that is commonly understood as the intersection between two different cultures—the Islander and the non-Islander, the latter expressed as Australian, Western, mainstream of whatever" (Nakata 2007, 322). In this space, standpoint theory is a way to recapture the power that mainstream society has taken away.

And this is the case in classification spaces too. By invoking the concept of positionality with reference to our indexed identities in classifications, we co-opt these spaces as ones that should necessarily empower us, and support and engender a sense of well-being in the world for each and every one of us. Classifications are just one part of the intersectionality that defines our lived reality. Evoking positionality is a move that can, hopefully, facilitate the liberation of these spaces and make them more representative, just, and participatory (Mathiesen 2016). Though organisms other than humans do not experience identity production in quite the same way, this move gives classification builders entrée into the general identity-producing mechanisms of these systems. Animals and botanicals do have certain rights to exist cooperatively in this world. Thus, we now turn our attention to the global nature of biodiversity classifications. Bit by bit, we will deconstruct and critique these digital ecologies such that we can get a sense of how we might approach the production of classifications that are mutually beneficial for all members of our social and natural worlds.

3 GLOBALITY

HYPOTHETICAL WEEVILS

Working with biodiversity classifications is not for the fainthearted, if analyzing seemingly insurmountable caches of data is not in your bailiwick. Biodiversity taxonomies are no longer defined solely by the local, which, by and large, defined taxonomic work prior to our contemporary, computationally ubiquitous period. If we look to the relatively recent history of biodiversity taxonomic science, one of the greatest changes to the discipline has been the globalization of information and the emergence of centralized aggregated spaces of taxonomic data. The reality is that taxonomies are now predominantly global, or at least it is the global taxonomies that are beginning to garner more attention. Spurred on by big data genome projects such as the Human Genome Project (Leonelli 2016, 17–18), the biological domains—and especially in the field known as computationally intensive bioinformatics (Stevens 2013)—required both an increase in the capacity to store and maintain data in centralized repositories and the coordinating mechanisms to share data across data-intensive projects. The same can be said for the biodiversity taxonomic world, which in the 1990s also began an infrastructural explosion of its own.

Well into the late 1980s and early 1990s, analog nomenclature lists—index listings of formal species names and their synonyms—were the most prevalent form of data used in taxonomic practice. For example, the printed Index of Fungi (Petrak 1969) was *the* official listing for fungus specialists well into the 1990s. When I first met Paul Kirk, then the main editor of Index Fungorum (2021)—the name for the now-digital version of the Index of Fungi—at his Kew office in London in 2017, he started our conversation by pointing to a row of books sitting on the shelf above his desk that collectively formed the core text for the Index Fungorum, which began as an initiative at the Centre for Agriculture and Biosciences International (CABI). When the International Mycology Institute, an initiative under the administration of CABI, integrated computational technologies into their work processes, Paul Kirk and colleagues meticulously transferred the Index of Fungi, a 625-page dictionary of fungal terms, as well as numerous index cards inherited from subscription-based index listing services prior to the 1980s, into digital form in the late 1980s and 1990s (Kirk 2017). As of October 2021, Index Fungorum holds 532,288 online records contributed by more than a thousand individual authors (Royal Botanic Gardens Kew, 2021), which represents both the historical core of the literature and the myriad "born-digital" contributions added to the database since its inception. Truly a project of great proportions, rife with details that would confound even the most patient information specialist.

Through the eyes of the average nonscientist, the result of biodiversity work such as this seems deceptively, even alluringly, simple. After all, it seems easy to list things—one need only briefly search an online search engine for the phrase "bird-watching therapy" to see how therapeutic and serene bird-watching can be! You just look for things *in nature* and voila, you begin making your list! Lists such as these, however, are compiled at varying levels complexity and their range of contributors is likewise surprisingly complex and varied. On the ground, so to speak, scientists will typically specialize in a family or genus of organisms. For example, a scientist or group might study weevils (order: Coleoptera) and compile detailed listings of species within that group. Nico Franz's (2020) work—a prominent systematist active in biodiversity informatics—studies the evolutionary history of weevils, with an even more specific focus on Neotropical and Sonoran Desert species groups. As one might imagine, the databases of information produced from these studies are often then contributed to a larger entity for integration into a larger cache of data. The road from a scientist's desk to the global scene, however, is neither clear nor simple to traverse.

To illustrate how local data might travel to global spaces, let us imagine a hypothetical weevil database for the Sonoran Desert—let's call it WeevilBase—containing species names, along with data related to each species. This supporting data could be most anything, including specimen measurements, qualitative descriptions, field site sample numbers, contributor and institutional information, genetic information, geographic coordinates, ecological context, images, and so on. Suppose it dawns on the builder of WeevilBase that this data might be useful for other scientists in the world, and so this taxonomist goes about finding possible locations to deposit their data. More than likely, in today's environment, a scientist interested in sharing data contributes their data to a database like the Global Biodiversity Information Facility (GBIF), arguably the most visible, and certainly the largest, aggregator of biodiversity data points on the planet. GBIF was formally established in 2001 as an open access environment, with the primary intention of aggregating biodiversity data toward the goal of a global view of extant biological knowledge.

GBIF functions through a distributive structure, with its Copenhagen-based secretariat supported by a governing board, as well as various standing committees and task groups populated by scientists and professionals around the world (GBIF 2020a). GBIF also has important policy-oriented roles in the global infrastructure of biodiversity data. The organization is charged with coordinating a series of global nodes (operational bases) throughout the world. These nodes are often located in a prominent natural history museum or other biodiversity-related institution within the node country. As of April 2021, one hundred such nodes had agreed to a nonbinding memorandum of understanding, promising to coordinate and maintain an internal institutional network of data creators. Contributors are from all over the globe, from Andorra to Zimbabwe, Argentina to Vietnam. Funding from GBIF typically comes directly from governmental agencies, large scientific institutions, and universities. Given this reality, GBIF has a prominent voice in both local and global polices that relate not only to data but also to the broader realm of biodiversity activities.

Returning to WeevilBase, let us imagine another scenario—one in which a scientist did not want to donate all supplemental data to an open-access repository, but rather wished to donate only species names. This is also quite valuable, of course. As in the Index Fungorum, valid, code-compliant names are an important organizing mechanism for databases, and provide vital information for scientists all over the globe. Species checklists and nomenclatures aim to be a "universal and complete" reference that identifies which species exist in a particular area; without this information, "we cannot sustainably use, explore, monitor, manage and protect biodiversity resources" (Species 2000 2017a). Species checklists are compiled for various reasons and function most effectively if they are "integrated, coordinated and disseminated from a single platform" (Hamer, Victor, and Smith 2012, 1). To aid in this broader system of names, then, the owner of WeevilBase, based (hypothetically) in the United States, has chosen to contribute this data to the Integrated Taxonomic Information System (ITIS 2020). Based at the Smithsonian National Museum of Natural History, ITIS is an important source of species names, especially built for governmental use. The names in ITIS are code-compliant, meaning they have followed international standards for their articulation and are accepted in the scientific community as being correct. These names are then embedded into a relational hierarchy that is tantamount to a database-style tree of life. This hierarchy—often called a taxonomic backbone, similar to the Catalogue's—is then used to organize data coming into an organization. The US Department of Agriculture, for example, might use this list to validate and control the possible importation of vectors into the country. Or, perhaps, certain subsectors of the US Environmental Protection Agency might use these taxonomies to manage incoming environmental monitoring data. By donating one of the best databases on weevils, WeevilBase's creator is greatly contributing to the proper management of weevils throughout the United States, including perhaps larger-scale ecological studies that rely on that same data to monitor fluctuating species counts in response to climate change.

The management of ecological matters in the United States is one thing, but the country-specific data is itself also valuable as part of a larger, global context. Biodiversity data is, at its heart, data about local phenomena that have global applicability and importance, but wedding local knowledge with global data markets is no easy task. Paul Edwards emphasizes this distinction, and the shift from localized to global data, in his *A Vast Machine*, an extensive study of climate data and organizations (2010, chaps. 8, 10). As Edwards emphasizes, standards and policies are absolute necessities to accomplish this kind of intercountry structural infrastructure. This process of standardization and normalization is a top-down process that requires data be examined and organized broadly, such that organizations all over the world can take advantage of resources. But these standardization mechanisms often ensure that local practices are diluted to cater to the lowest common data denominator.

If one were to provide a list of the most pertinent databases in the global biodiversity and biological taxonomic world today, one would list a bevy of acronyms and distributed initiatives on top of what we've already discussed: International Barcode of Life (BoL), World Register of Marine Mammals (WoRMS), the Biodiversity Heritage Library (BHL), and the Encyclopedia of Life (EoL), to name just a few. The fact is that, like the domain of physics and climate work, biodiversity work has become increasingly global and distributed. The organizational lattice needed to support, maintain, and make accessible biodiversity information is not uniform or singular, but is rather a network of organizations that work collaboratively to maintain some of the largest representational infrastructures in science. Within any network, you have organizations specializing in certain tasks, which might include designing an organizational structure to keep data uniform (such as ITIS), collecting and maintaining data (such as GBIF), providing access to historical literature (such as BHL), providing access to vital knowledge resources (EoL), and advocating for policies related to data (GBIF). To activate the global potential of WeevilBase data, ITIS would have to have clear mechanisms to engage with this global data consortium.

Enter, the Catalogue of Life.

THE CATALOGUE OF LIFE

Luckily for WeevilBase, ITIS is a core partner of the Catalogue of Life, which as introduced in Chapter 1 is one of the globe's principal taxonomic backbones for biodiversity data. The Catalogue drew a great deal of attention within the scientific community when it formed in 2001, mostly from scientists quite optimistic about the prospect of an aggregative, authoritative taxonomic system (Reichhardt 1999; Bisby et al. 2002; Cachuela-Palacio 2006; Gewin 2002). As Frank Bisby outlines in his article, "The Quiet Revolution: Biodiversity Informatics and the Internet" (2000), early organizations like Species 2000 (2015a)—a joint program between the European-based International Council for Science: Committee on Data for Science and Technology; the International Union of Biological Sciences; and the International Union of Microbiological Societies—initiated a collaborative mechanism to produce backbone taxonomies to support the aggregation of data from disparate entities around the globe. The Catalogue is one result of this push. The Catalogue comprises two previously independent entities that merged in June 2001 (Bisby et al. 2002): the Species 2000 organization (2015b), which covers species across the globe (with an original emphasis on European species), and ITIS.

The primary motivating individual for this pan-Atlantic merger, Frank Bisby, believed that the globalization of biodiversity data, and the interoperability this centralization facilitated, was essential to expanding the capacities of future research in the biodiversity sciences (Bisby 2000). Species 2000 remains the current legal body for the Catalogue, charged with the responsibility of coordinating contributing entities and governing the use and policies of the produced data (Species 2000 Secretariat 2015a). The Catalogue has two core functions significant to this narrative: (1) it has charged itself with compiling the most comprehensive listing of all known existing species on the planet, and (2) it arranges these species lists into a consensus-based classification that can be subsequently used to organize the data for its partner institutions.

As of October 2021, the Catalogue contained more than 2 million species, populated by more than 163 individual databases from around the globe (Species 2000 2021b). Presuming a total of 2.3 million extent species, the Catalogue has successfully mapped approximately 87 percent of life on the planet (Species 2000 2021c).[1] While ITIS's total contributions initially made up 50 percent of the Catalogue's Annual Checklist, their percentage has decreased over time. Though detailed data from the 2000–2004 data sets is no longer easily available, contributions from ITIS constituted 158,884 of the 220,000 core species names in the taxonomic Catalogue of Life database in 2005—a full 72 percent of the total database species count. Comparing these figures with more recent data from the 2019 release, ITIS contributions have increased by only a small margin, to 148,975 total species. However, given the Catalogue species total is upward of 2 million species and still growing, ITIS is contributing a smaller and smaller percentage of the total database environment over time. The diminishing role of ITIS in the Catalogue's totals points to a rather significant fact: as more specialized databases enter the system, we must contend with the concomitant complexities this adds to the system. Species lists are notorious for continually changing, for reasons that will become obvious, but suffice to say for the moment that stability is a rare quality of these lists. But if the Catalogue is intended to be a system to communicate the evolving landscape of biodiversity data, a question arises as to how best to deliver information in a way that properly expresses this dynamic reality. During a Catalogue of Life meeting in 2017, a conversation precipitated among a group of scientists discussing the implications of making nomenclatural data sets available before being edited:

Participant 1: Most of the researchers consider this database a work "in progress." It's never finished. This is part of the problem. It's that the work is not finished.

Participant 2: I'm afraid to open raw data to the public because somebody will take the data without knowing what happened in the work bench. It *is* eventually finished, but it doesn't loop back. The Catalogue of Life encourages the publishing of draft systems monthly. [The] final, annual checklist will have a more polished presentation.

Participant 3: There is a psychological element. Scientists want it to be perfect but it never will be.

Participant 4: Users don't see how much work goes into compiling the database.

Participant 1: Because most of the custodians do it in their spare time. If you have an incomplete data set you get questions and you have to spend time answering them.

Participant 5: If you care about the science you don't want to publish something that isn't refined. We strive for perfection because we have the knowledge and want to pass it on. We feel a disservice if it is not finished.

As can be gleaned from this discussion, the when and how best to publish data is an ongoing issue of concern, coupled with the reality that taxonomic inference is always, to a certain extent, ongoing and not fully "completed" in the proper sense of the term. As Thiele and Yeates (2002) penned in *Nature*, the hypotheses represented by biodiversity taxonomies are particularly volatile, making consensus, properly speaking, a fleeting target. There are always more specimens to collect and more opportunities to optimally refine data. Additionally, since the Catalogue ingests taxonomies as they become available and staggers database updates over the course of the year, a portion of the Catalogue is always in "process," so to speak—even beyond the ongoing taxonomic work experienced at the local level.

To offset this unavoidable issue, a two-pronged publishing model has been established. As indicated by then executive editor for the Catalogue Yury Roskov, based at the University of Illinois at Urbana-Champaign at the time of publication, the Catalogue has begun to see itself function under the serials model of production:

> My contribution to the aggregation and editorial process was identifying that we need to move the Catalogue of Life as close as possible to the traditional way of scientific journals. It means that if you have a choice [among] different taxonomic databases that cover the same group, we need to have a peer review process where independent reviewers will tell us which is the best source, and this takes time. And so we established the monthly and annual editions (2016a).

The first iteration is a "dynamic" monthly edition of the database, which is meant to provide an easily searchable up-to-date snapshot of Catalogue data. What you lose with this method currently, however, is a product that has been carefully vetted and approved by the editorial process. This edition is not archived, per se, but is rather meant to express as accurately as possible the iterative nature of the database. The documentation associated with this version is limited: "Anything can change as the [species] list develops: names, their associated details, and their content providers—and there is no tracking of those changes. For that reason, the monthly edition is not the one to quote if you wish to cite a verifiable source" (the ephemerality of these data sets is indicated by the dotted boxes in figure 3.1) (Species 2000 2015d).

The second iteration, the Annual Checklist editions, are published both online and in CD form in what the Catalogue calls "fixed imprints" (Species 2000 2017b). This version is static, citable, and formally published and identified through an International Standard Serial Number and an edition designation (an identification number applied to journals and other serial publications). Figure 3.1 outlines the products and entry points for the annual version of the Catalogue. This is the version that is often implemented as the backbone structure for other information products. The annual version also provides the opportunity to compare iterative data sets and is still published in compact disc form (many developing countries, lacking high-speed internet, still prefer or require disc format, despite its fall from favor elsewhere). As Geoffrey Bowker (2008) has articulated, databases contain the root elements of a narrative that are vital to understanding the constitution of social and scientific practices. So, the annual narrative serves as a mechanism for better understating the development of the Catalogues "consensus" over time. Synonymy fields are built into the Catalogue to allow for variable terminology for species—an issue that remains common in most classification systems, including the Catalogue.

With the nomenclature as the base, the Catalogue then serves a second function: it relates these species lists into hierarchies that serve as a backbone taxonomy for databases throughout the world. The "management classification," as it is called, is curated by a series of experts and, unlike other systems, sidesteps the use of algorithms to instead find the best arrangement, given the innate conflicts among different taxonomic opinions (Species 2000 2015b). Contributed data sets are distributed among

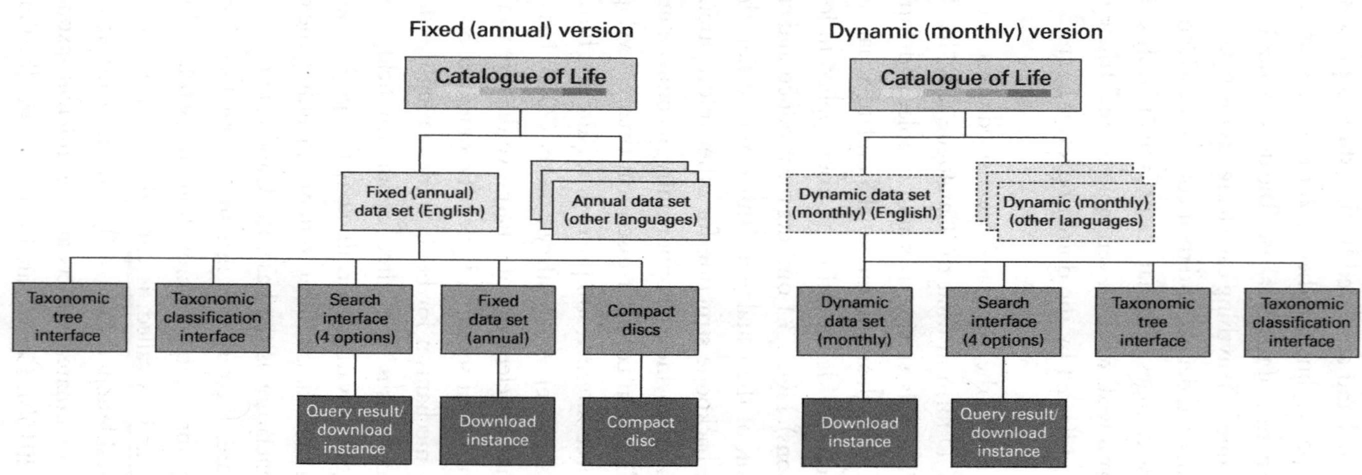

Figure 3.1
Publishing model for the Catalogue of Life. The left flow chart indicates the entity types for the fixed (annual) version of the Catalogue. The right flow chart indicates the entities for the dynamic (monthly) version of the Catalogue. The dynamic version is not archived or saved for later use, so they are temporary exemplar documents (indicated by dotted lines).

three databases types in the Catalogue: global species database (GSD), regional species database (RSD), and a thematic species database (TSD) (Species 2000 2014). GSDs contain worldwide coverage of the species within taxon (all the weevils in the world compiled in one space). RSDs contain regional coverage of a species within one taxonomic group (all of the weevils from the Sonoran Desert, such as our WeevilBase). TSDs are particular arrangements of species collected for reasons other than primarily geographic coverage (weevils that have been artificially introduced to a certain area, for example, or weevil species that live only in caves on the European continent). These subsidiary databases are contributed from various locations around the world, including from the likes of Royal Botanic Gardens, Kew; the World Register of Marine Species (WoRMS) (2017); Fishbase (2017); and Systema Dipterorum (Pape and Thompson 2017), which constitute the largest subsidiary databases within the Catalogue (Species 2000 2015b). Over time, as more and more GSDs and RSDs are added the Catalogue, species checklists become more robust. The Catalogue stands at the center of a multitiered infrastructure, ingesting subsidiary databases from regional hubs from around the world (see figure 3.2).

An important issue to reiterate is that no universally accepted reference taxonomy currently exists in the biodiversity world. As such, conflicts between contributed taxonomies within composite structures do exist in abundance. Recall that our invented regional species database, WeevilBase, was compiled by one scientist, and therefore the taxonomic arrangement of weevils as it exists in that database represents one scientific opinion. Let's assume that WeevilBase was compiled with phenetic commitments, meaning that species relationships were based primarily on physical and morphological characteristics. Now imagine that another global species database for the order Coleoptera, hypothetically named BeetleMania, which comprises all beetle species, including weevils, was also accepted for inclusion in the Catalogue. BeetleMania, however, uses a cladistic approach, meaning that species are arranged based on most recent ancestor relationships. The result is that WeevilBase and BeetleMania will conflict at the level of weevils in terms of their hierarchy. Editorial intervention by the Catalogue is then needed to reconcile these differences.

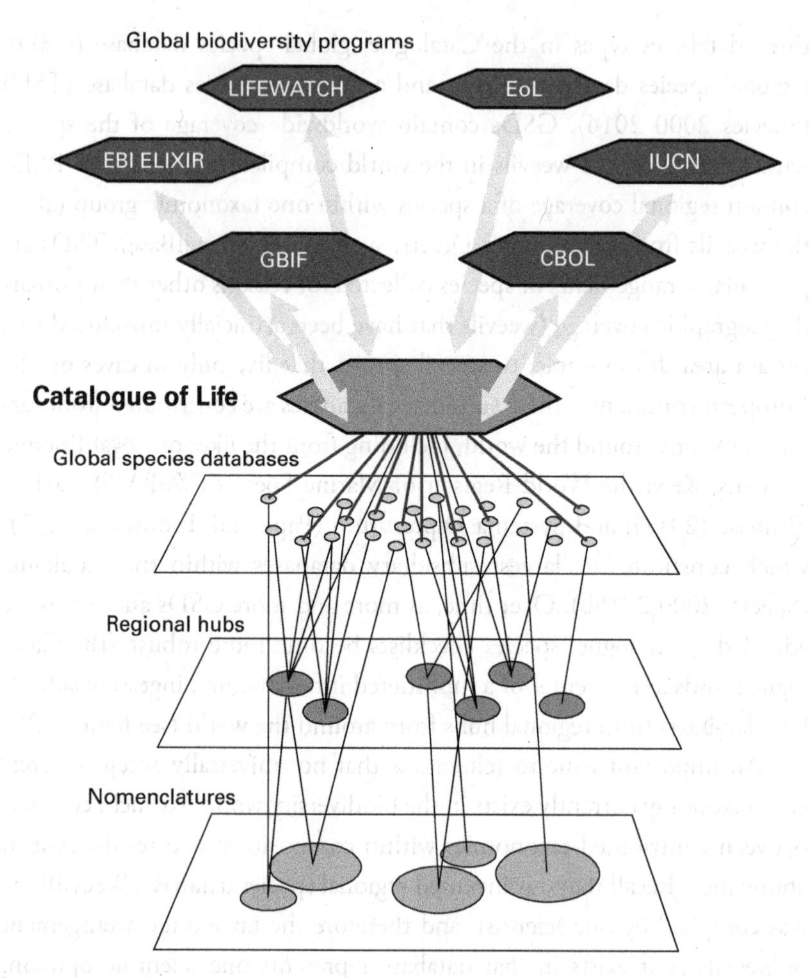

Figure 3.2

Catalogue of Life infrastructure layers (Species 2000 2015a). Nomenclatures exist at the bottom of the infrastructure layer and include all code-governed nomenclatural acts, including original name usages (in taxonomic literature) and objective synonyms, as well as other name forms. Regional hubs are regional checklists (RSDs) for a given geographic area. Global species databases (GSD) give taxa on a global scale. The Catalogue is then used as a taxonomic backbone for many other online systems. CC-BY 4.0, Catalog of Life, used by permission.

In the scenario above, WeevilBase, which began as a small, regionally based database, has now successfully radiated throughout the biodiversity data network. The result of this move toward composite structures is of great consequence to the way we understand the epistemic qualities of classifications, particularly because, as more and more subsidiary databases come together to create a management classification scheme, it becomes increasingly more difficult to differentiate the permutations exacted on the data and hierarchical structures in any given taxonomic database. In the process of this dislocation, data is wrested from its original context. In the process of gaining global authority, we lose aspects of the data's local integrity.

This process of composition, then, wields great organizational power within the field of biodiversity taxonomy as whole. And as Andersen and Skouvig have shown with their collection of essays, *The Organization of Knowledge: Caught between Global Structures and Local Meaning*, "Contemporary digital information society has globalized information structures and facilitated easier access to information across libraries, cultural institutions, and in the Internet. While this has helped shaped global discourses it has often done so at the expense of localized meaning and ethics" (2017, xi–xv). Obfuscating the local in deference to the global brings about its own professional and epistemic challenges.

Miranda Fricker (2009) makes a similar note about the systemic qualities of the distribution of power in *Epistemic Injustice: Power and the Ethics of Knowing*. In Fricker's context, a key notion of social power is that it is structurally embedded; to offset the epistemic injustices that precipitate from within any society, we must be able to trace how these myriad effects are caused by multiform, intersecting mechanisms. Although power can be exerted individually (as in person to person), perhaps its strongest trait is that the levers of injustice are invisible and impossible to pinpoint in any one location. In distributed classificatory spaces, of which the Catalogue is just one, this obfuscation of power is subsumed within a complex arrangement of standards, tools, policies, and managerial decisions. Which reiterates Foucault's point: "Power in the substantive sense, *le' pouvoir*, doesn't exist." Foucault notes,

> What I mean is this. The idea that there is either located at—or emanating from—a given point something which is a "power" seems to me to be based

> on a misguided analysis, one which at all events fails to account for a considerable number of phenomena. In reality power means relations, a more-or-less organized, hierarchical, co-ordinated cluster of relations. (Foucault and Gordon 1980, 198–199)

Understanding how to trace the routes of power within institutions is quite difficult, to say the least. Foucault said, "Truth is to be understood as a system of ordered procedures for the production, regulation, distribution, circulation and operation of statements" (Foucault and Gordon 1980, 133). A virtue of a system like the CoL is that it absorbs the burden of the hefty task of data aggregation in ways that, for at least some, productively brings local knowledge in conversation with global knowledge systems. One downside, however, is that aggregation masks complex intellectual operations that dictate the usefulness of the system to any one scientist or user. It is difficult enough to wrap your head around the organizational complexity of figure 3.2, and this figure only hints at the many epistemically significant decisions occurring at each of these levels. As T. S. Eliot notes, however, "In a minute there is time / For decisions and revisions which a minute will reverse" (Eliot and Carr 2002). How these decisions will come to be understood and contextualized in the larger CoL system is the burning question in the minds of many taxonomists.

This isn't to say that any one individual within the Catalogue is necessarily responsible for any particular obfuscation that might occur. This phenomenon is a practical quality of all systems, within and without biodiversity taxonomic work. As Bowker and Starr wrote, "Once a system is in place, the practical politics of these [classification] decisions are often forgotten, literally buried in the archives (when records are kept at all) or built into the software of the sizes and composition of things" (1999, 45). Iris Marion Young similarly notes, in *Responsibility for Justice* (2011, xviii), that while an individual might ultimately be the agent enacting the injustices facilitated within systems, the embedded, structural components of systems facilitate injustices in ways that extend far beyond individual responsibility. She speaks of a woman named Sandy, forced to relocate from her dilapidated condo, who begins to seek affordable housing for herself and her children. Unable to meet the expensive demands of rent, a down payment, and car payment, Sandy is forced to question whether homelessness might be in her future. Young's point is that, while any *one* of these variables might independently seem tenable, if unfairly inflated, it is the intersection of these high prices, embedded as they are within a larger, exploitative economic system that creates the environment for her unjust position. Similarly, while the Catalogue might be faulted (rightly or wrongly) for many of the decisions implemented in their system, the reality is that, given the complexity of taxonomy work in general, epistemic obfuscation *of some kind* is inevitable. One cannot "archive" every step of the decision-making process, whether at the local, individual, or global level.

SOCIAL-ECOLOGICAL SYSTEMS

Structurally, the Catalogue is a complex open-source system that must meet the current needs of biodiversity specialists. As will be discussed, it also provides data to prognosticate future needs for data that are not yet at the forefront of asset discovery and research. In thinking about how to critique the circulation of power within systems of this complexity, I have found it helpful to think in terms of Elinor Ostrom's, "A General Framework for Analyzing Sustainability of Social-Ecological Systems" (2009). As a political economist, Ostrom's primary interest in this piece is identifying a structural analytic by which we can understand how best organizations (governments, industries, and the like) can *sustainably* yield resources from the natural world. In the article, Ostrom focuses on the lobster fishing industry of the coast of Maine, and notes that many subsystems within this area merit specific attention: "resource systems (coastal fishery), resources units (lobsters), user (fishers), and governance systems (organizations and rules that govern fishing on that coast)" (2009, 419). A central problem in this approach for Ostrom, however, is that in order to effectively manage social-ecological systems (SES) that are sustainable, one needs to look *across* disciplines—for example, to knowledge from the social sciences and ecological sciences, which operate under different sets of epistemic and methodological assumptions. "Thus," Ostrom writes, "we must learn how to dissect and harness complexity, rather than eliminate it from systems" (2009, 420).

In light of this, Ostrom creates a detailed framework that breaks down SES into first-order variables and second-order variables that can be used to assess their suitability and efficacy. Each component within this framework is seen to intersect with another, as well as with the social, ecological, and political contexts of that SES (2009, p. 420). Ostrom's subsystems include a resource system; its resource units; its governance systems; and its users. Each subsystem is then broken down into its component variables that, comprehensively, can be used to examine how sustainable that industry system might be, both in the present and as it proceeds into the future. So, if we were to take the Catalogue of Life as our resource system, the variable Ostrom (2009, 421) identifies as significant would be,

RS1 Sector (e.g., water, forests, pasture, fish)
RS2 Clarity of system boundaries
RS3 Size of resource system
RS4 Human-constructed facilities
RS5 Productivity of system
RS6 Equilibrium properties
RS7 Predictability of system dynamics*
RS8 Storage characteristics
RS9 Location

In this case, since Ostrom is concerned with the sustainability of natural resources within certain sectors of economic activity, the application of this model has its limits when we look to the mechanisms of power within an entity like the Catalogue. However, I argue that, with some tweaking, Ostrom's subsystem components could be amended in ways that more closely align with the concerns of a classificatory environment.

To avoid the trap of getting too far ahead of our ourselves, let us focus on just the first three variables within the resource system (RS) category. Let us ask this question: In what ways is the Catalogue suppressing the transparency of local knowledge in its global aggregated system? The sector

(RS1) variable is straightforward enough: biodiversity taxonomic work—so far, so good. Clarity of system boundaries (RS2) is interesting, however, particularly because, as classifications are brought into the management taxonomy and set beside each other in the CoL (see figure 3.2), we aren't entirely sure where the taxonomic principles of the *contributed* taxonomies end and the principles of the *management* taxonomy of the Catalogue begin. In this way, the clarity of taxonomic boundaries of local knowledge are somewhat hazy. Looking to the size of the resource system (RS3), we might also make the claim that the size of the resource system is both too broad *and* too limited. It might be too broad, given that the larger the Catalogue gets, the more diluted local knowledge might become. On the other hand, the system might be too narrow, given that it accepts only knowledge structures specific to Western scientific sensibilities and ignores (perhaps understandably) diverse, counter-epistemic knowledge systems of indigenous tribes throughout the world. Surely, indigenous knowledge has something to add to this global picture. (We examine the problematic of the Western gaze in chapter 8.)

Theoretically, we can proceed with Ostrom's analytic and break down the various ways local knowledge might be suppressed and obfuscated by way of the Catalogue's overall structure. The function of this chapter has been to illustrate how locally derived species and taxonomic information might find its way into global infrastructures. In the brief examination of WeevilBase, we can begin to imagine the levels of complexity introduced at each step throughout the process: from the Sonoran Desert, where local data on weevils is collected; to the desk of the weevil scientist, where taxonomic charts are carefully constructed; to the global database space of GBIF, where WeevilBase data becomes intricately enmeshed with data derived by scientists throughout the globe. The gist of the matter is that, although the discipline of information studies has posited power as a central mechanism of concern in the space of classification and representation, the discipline has yet to see many deliberate deconstructions of systems in the manner proposed by Ostrom. I posit this book as an example of how we might use a model such as Ostrom's to locate power levers within classifications and thereby better understand their cultural impacts on our society.

4 CONSTRUCTING TAXON CONCEPTS

> An organizational scheme might choose any features at all of the writings as the basis for assignment to position: size of copies, weight of copies, security classification, legal or social function, religious orientation. But of the familiar sorts of organizational scheme, those resembling The Catalog in assigning positions on the basis of subject matter are perhaps the most interesting and problematic.
>
> —PATRICK WILSON
>
> *Two Kinds of Power: An Essay on Bibliographical Control* (1968, 69)

ON CONCEPTS AND ARTIFICIALITY

One of the greatest powers of the classifier is, quite literally, to make concepts and entities *exist* within a system—to instantiate concepts as tools for description and conduits for information access. To classify is to have the *power to* create a concept within a system, as well as the *power over* how people both conceptualize that system and subsequently use that system to access resources. Whether the dingo exists in a classification as a genuine species of its own dictates the way Australian politicians and governments can act in response to whether it is seen as a pest in the country or not. But what, exactly, does it mean when I say, "Classifications bring entities into existence?" Surely, the organism we associate with being dingo-like existed before the name *Canis lupus dingo* or *Canis familiaris dingo* was formally entered as a classification. Just as entities in the natural world existed before Linnaeus articulated the formal rules for the application of binomial nomenclature.

The point of this question reaches to the heart of what is meant when I say that classifications are constructed: all the concepts and entities that we include in classifications are socially produced, even if their articulation is part of an empirically based scientific process. To say that there are no natural categories is to claim that all attempts to fragment the world into disparate entities is, and always will be, an artificial activity. Classification may be a natural human inclination, yes, but the classes we produce are a product of our own cognitive and disciplinary limitations, spatiotemporally contingent interpretations, and methodological possibilities. "Living systems must categorize. Since we are neural beings, our categories are formed through our embodiment. What that means is that categories are a part of our experience! . . . We cannot . . . 'get beyond' our categories and have a purely uncategorized and unconceptualized experience" (Lakoff and Johnson 2010, 19). Categorization and classification are always performed relative to one's self and the society in which they are embedded. Even the "natural" cannot exist on its own, for to understand what is natural we must relate it to what is not natural—namely, everything that humanity has created in an artificial sense, such as cities, towns, and other human-made environs. Additionally, these classification systems inherit the just and unjust elements of that same society from which they emerge. There is no foundational rubric on how to distinguish one species of giraffe from another, just as there are no preestablished guidelines on how we should categorize one's sexual preferences, ethnicity, gender identity, or, far more innocuously, various shades of blue or red. How we set the boundaries for these classes is a sociocultural issue.

Importantly, a crucial historical distinction has been made between what biodiversity taxonomists have considered "natural" and "artificial" systems. So, while much of this discussion is in service to IS scholars perhaps unacquainted with biological taxonomic history, the reality is that working biological taxonomists are well aware of the artificiality of their constructions. According to Staffan Müller-Wille, the distinction between natural and artificial systems first arose when the distinction was identified by Carl Linnaeus (Müller-Wille 2013), though Phillip Sloan (1972) traces the genesis of this juxtaposition to the seventeenth century, especially by way of the writings of John Ray, Joseph Pitton de Tounefort, and August Bachmann. In a general sense, the articulation of a natural system has, as noted by Phillip Sloan, attempted to devise a method of taxonomic construction that "neither reflects whimsy of the taxonomist, nor represents simply a utilitarian cataloguing device like the Dewey Decimal system" (1972, 2). How any one scientist defines what constitutes an appropriate method to devise a natural system has varied by time period—conservatively ranging from methods that identify a single character, or finely limited set of characters, as viable candidates to inform the production of a natural system (the fruit-producing parts of plants, for example) to whether a great many characters should be considered for the construction of such a system (such as in phenetical taxonomy). As Müller-Wille notes, natural systems, at least since the time of Linnaeus, can more broadly be defined as those that "conform to intuitions about group membership that are based on overall resemblance" (Müller-Wille 2013, 310), using a variety of characteristics to devise a comprehensive comparative study. The goal for defining a natural system is, at least in part, to locate an inherent natural, intrinsic order by empirical means (Lefèvre, 193), however that 'intrinsic nature' might be defined at a given time period. As also pointed out by Sloan, how this natural ordering has been interpreted (that is to say, what it has represented) has differed based on historical period. For pre-Darwinian taxonomists, a natural system was one that mirrored the "true essence" of the natural world (Sloan 1972, 2), inclusive of arrangements such as the Great Chain of Being, which held God, angels, and humans at the apex of the organism- and mineral-based hierarchy. In a post-Darwinistic world, a natural system came to reflect a phylogenetic framework that was most likely accurate to evolutionary history—a belief that still is, in many ways, latent in some of the discourse surrounding phylogenetic classificatory arrangements (Doolittle 1999a; Liu, He, and Schneider 2014).

As noted by Stefan Müller-Wille (2013), it was Linnaeus who first used the terms natural and artificial to distinguish these concepts; to foist this distinction squarely over disagreements prior to Linnaeus's time is to fall into the danger of anachronism and oversimplification of the disagreement. Linnaeus believed that his own system, as well as all other taxonomies devised up to that point, were definitionally artificial, and that to have any chance at devising a natural system would entail a near full catalogue of organism characteristics (Müller-Wille 2013)—a feat unattainable in the eighteenth century and perhaps debatably an impossible task even in the present or at any point in the future. Artificial systems, then, are context-dependent taxonomies, based on the local and practical conditions under which these taxonomies are produced. These classifications are based on limited character comparisons and necessarily produce pragmatic classifications. In an ideal world, a natural system would require no great revisions upon the addition of a new organism. After all, a natural system would be comprehensive enough to fit all current and future organismic discoveries. Artificial systems, however, are often dependent on the select characteristics and subject to great change upon the addition of new species—for example, particularly when new specimens introduce ambiguity with respect to the characteristics used as the principal divisions for a taxonomic hierarchy (Müller-Wille 2013, 311).

So, it is one thing to speak of the arbitrary nature of classification in a general sense—as in Sloan's indication that the Dewey Decimal Classification is utilitarian—but it is quite another to transport those assumptions full-scale to the biodiversity classification world. My use of the word artificial may gesture to this historical contention, but in general, my aim is to take this artificiality as a given (as most all biodiversity taxonomists do) and to better understand how the conditions under which classifications are constructed produce certain modes of systemic imbalances of power. We know that taxonomic practice is anything but arbitrary and that the construction of biological classifications is based on careful scientific analysis and methodological inference. Much like the concept of literary warrant in information studies (Beghtol 1986), the construction of a new class (bibliographical subject or taxon, for example) requires the establishment of evidence that justifies the need for a new class. To say they are artificial is not a slight, but rather a reality of classificatory construction.

SPECIES CONCEPTS

With this distinction established, two terms merit definition before we proceed onward: species concepts and species taxon concepts, which Walter Bock (2004) notes are too often confused in biological and systematic literature. First, to the species concept: as defined by Bock (2004), the most widely recognized is the biological species concept (BSC) or genetical species concept (GSC). The GSC is defined as a "group of actually or potentially interbreeding populations which are *genetically* isolated in nature from such other groups" (Bock 2004, 180; emphasis original) The BSC, originally identified by Ernest Mayr, is slightly narrower, noting that a species comprises a population that is reproductively isolated. But as Bock has noted, some species can interbreed without producing viable offspring; for example, a horse and a donkey can interbreed to produce a mule, but the mule is infertile because of chromosomal differences between the two parent species. It is the case that genetic isolation is concomitant with reproductive isolation, but in this case, the primary determining factor is genetic. It is not always the case that reproductive isolation equals genetic isolation. In addition to these concepts, as summarized by Marc Ereshefsky (Ereshefsky 2007, chap. 2), quite a few other species concepts are also in circulation—some say upward of twenty-two to twenty-six, in fact (Wilkins 2011). Each species concept has its own theoretical approach to defining how organisms can be meaningfully and functionally grouped into evolutionary units. For example, a phenetic species concept is one that defines a group of organisms based on their resemblances with each other. This definition is operational "based only on observable facts of similarity and discontinuity" (Winston 1999, 44). Here, one might say, after a close morphological examination of our aforementioned giraffes, they are one species because they are similar enough to one another that such a determination makes good empirical sense. Whether they can reproduce with each other might, for some, solidify such a categorization, but reproduction would not make or break a phenetic determination. Another species concept is the ecological species concept, which defines a species based on the environmental niche they fill and the extent to which a group utilizes a particular set of natural resources in an ecologically efficient way. Such an approach prioritizes environmental forces and conditions as the stabilizing force for a species, over and above reproductive isolation (Ereshefsky 2007, 87–90). In this way, the operative variable in the ecological species concept is the environment, even if interbreeding is also a quality of a group.

It is imperative to note that the species concept is a term that applies to the domain of evolutionary biology; its articulation is separate from the practices that taxonomists undergo creating species categories (a given level in the Linnaean hierarchy) and the species taxon, which is delimited (described) (Bock 2004). A working taxonomist may adopt any one species concept, and the subsequent articulation of species categories and species taxa will be influenced by that choice. It is often the case that one's research context determines the best species concept to implement in practice. As Kevin De Queiroz indicates, "Various properties [of species concepts] are of greatest interest to different subgroups of biologists. For example, reproductive incompatibilities are of central importance to biologists who study hybrid zones, niche differences are paramount for ecologists, and diagnosability and monophyly are fundamental for systematists" (2007, 880). Regardless of which species concept is adopted, the concept must then be consistently applied to all species taxa within a given hierarchy (Bock 2004, 185). Adopting different species concepts in the same taxonomic construction will inevitably lead to contradictory, and possibly incompatible, species categories and species taxon (De Queiroz 2007, 879–880). For one, some concepts facilitate the creation of more taxa (i.e., genetical or phylogenetical species concept; colloquially, a "splitter"), whereas another species concept might create taxa inclusive of more organisms, and thus fewer taxa (i.e., biological species concept; a "lumper") (2007, 879–880).

TAXON CONCEPTS

Taxon categories are the basic building blocks for the Linnaean hierarchy and species taxa in particular are the fundamental and foundational taxon concept on which all other taxa and hierarchy categories are defined (De Queiroz 2007, 181). Species concepts (as described above) inform the conceptualization of taxon categories in general, and species taxa in specific, by influencing how a particular species taxon is delimited (bounded), described, and related to other taxa. Taxon concepts are empirical formulations in the strongest sense of the word and function as any other scientific opinion might—as a hypothesis, subject to future testing and revision (Wiley and Lieberman 2011, 29). For example, the articulation of the species taxon concept and the delimitation for *Ursus arctos* (the concept commonly known as the brown or grizzly bear) can be challenged at any point by reformulating the taxonomic circumscription and concept through a new published delineation. Summed up by Sterner, Witteveen, and Franz,

> Unlike the choice of a species concept, which given the current state of the species debate can perhaps be treated as somewhat arbitrary, taxonomic concepts have an empirical status as scientific hypotheses. To see this, consider the difference between identifying what a given name refers to with and without specifying a taxonomic concept. It is easy and uncontentious to say, "This taxonomic name refers to the biological entity that includes this type specimen." While correct as a matter of principle, a statement of this kind fails to communicate anything about the present state of knowledge about the relevant species. It is much harder epistemically to accurately identify and agree on which organisms other than the type specimen are also members of the designated species. (2020, 8)

Taxon concepts are *definitional* concepts, in that a taxon represents a particular taxonomist's informed opinion about what constitutes a bounded population of real-life organismic units as observed in some natural environment (De Queiroz 2007, 182), ultimately represented by species nomenclature.

A taxon concept is complex, represented as it is within a computational system by a name that references numerous external sources, including the type specimen that anchors the name (Patterson et al. 2006). In Ronald Day's terms, "the name points to the object and the name reflects the networks in which the object first appears as a named thing" (Briet 1951, 49). The ties that bind these disparate sources of evidence, however, are shifting and difficult to trace as changes progress over time, such as a revision to, or a shift in the articulation of, a particular taxon concept. A species taxon concept in an operational sense is an accumulation of a name string in a technical system, as well as the compilation of referenced evidentiary objects that make that name valid as part of scientific discourse (the specimens used for description, including a type), as well as a publication that delimits and describes the taxon. Without this "trinity" in place (the type, the description or delimitation, and the taxon's stable nomenclature) a concept fails to make it into formal scientific discourse. Typification involves the identification of a type specimen—an individual or group of specimens. Types are typically held in museum repositories and act as the material ground for and the name-bearer of the species (Daston 2004). Description occurs through a publication act and includes a species diagnosis, a description, a taxonomic discussion, and, potentially, notes about a species taxon's ecological characteristics, geographic distribution, and other matters (Winston 1999). Instantiating a species taxon also involves attaching said delimitation to a name, which is of special interest in this discussion since names are the primary collocating mechanism for data about a species once taxon concepts are brought into a taxonomic space.

The Type Specimen

Most have a general idea that one of the primary documents of biological taxonomy is the type specimen—the one and most obvious physical thing that is "attached" to a name, that creates the species "out there." Michael Buckland's ruminations on museum specimens-as-documents (2017) and Suzanne Briet's (1951) trope of the antelope have permeated IS and documentation studies such that the type is often seen as the most vital taxon-representing object. Briet is well known for popularizing one of the most well-known definitions of the "document" in the field of IS and documentation studies: "a document is a proof in support of a fact" (1951, 9). To illustrate the complicated indexical relationship between documents and their referent, she uses the example of an antelope in the wild, which is then captured and brought back to Europe, with its ultimate fate being "stuffed and preserved in a museum" (1951). In the wild, the antelope is merely a natural object; but in a botanical garden in Europe or as a specimen in a museum, however, the once free and live antelope is now a document that indexes the "kind" of the antelope that lives free. Briet also describes many other documents that are created in this process (articles about the antelope, photographs, etc.). Many scholars in IS and documentation studies use these distinctions as a jumping-off point for theoretical examinations. The truth is that the type specimen is not related to the taxon concept as directly as we might think—and certainly not as neat and straightforward as Briet's exemplar. But this assumption subsists for good reason. After all, when establishing whether a given organism is a new species or a new discovery, one often, as Judith E. Winston states, must look to this physical object as evidence for taxa determinations (1999, 96). But while types do occasionally settle taxonomic disputes (Godfray 2007) and can also act as the voucher specimens on which some molecular examination of species are based (Seberg et al. 2016; Smithsonian Institution 2017), they are not always determinative. The most important function for type specimens remains to anchor and regulate nomenclature by preventing the chaotic inflation of names (Witteveen 2015, 570; Winston 1999, chap. 9). With this stability, the name, and its associated taxon concept, can then be argued in the space of scientific activity.

When Linnaeus began using binomial nomenclature, he based his description on specimen samples, but the formal identification of types was not yet a global standard (Ereshefsky 2007). Not until the late nineteenth century did typification enter common usage, followed thereafter by its formal integration into the codes of nomenclature (Daston 2004, 154–158). Before such rules were established in codes, the assumption that the type *be* typical led to an inflation of the application of types, as well the frequent replacement of types for newer exemplars that more adequately matched phenotypic observations (Hull 1988, 498). New names arose everywhere, many for the same species, leading taxonomists to mandate the use of types. Only one type could be applied to one name; by doing so, the production of names became more centralized and controlled (1988, 498), focusing efforts more on the process of taxon circumscription. The result of this codification is that, regardless of how a taxon is described over time, it remains what Joeri Witteveen calls a "necessary truth that the taxon's type specimen falls within its boundaries" (2015, 569).

And while the type is presumed to be "in one's mind" while describing and delimiting a species taxon, in reality, the type is not a "typical" representation of the taxon it is eventually associated with. One particular individual of a species can potentially vary descriptively from any other individual—males might look different from female specimens, and specimens might look different based on life-stage differences or natural variability. In general, best practices have been established for identifying a type. Ideally, when possible, a type should be selected from a robust and large population where variation is likely most ideal (Winston 1999, 177). In most cases, one specimen is identified as the primary name-bearing document; this specimen is called the holotype, which means a name is *always* directly tied to that one specimen. In addition to the holotype, paratypes can be identified, which are specimen references for description and delineation, but which are not in any way connected to the instantiation of a name (ICZN 1999, sec. 73). If many types of equal stature are used to describe a species, they are collectively known as syntypes, though this practice is certainly less favored within institutions, given the complexity of preserving types and managing their collocation in a collection.

Types can be of full or partial organisms, or some fossilized remains or impression of a previously extant species (ICZN 1999, sec. 72.5). As Rogers et al. note, "specimens are testable, tangible, and verifiable data sources" (2017, 456). Photographs proper cannot generally serve as the type without a great deal of scrutiny, as least not in the ICZN (1999, sec. 72.5.6). Though it is possible to base a species description on a photograph, description, or illustration of a type specimen, the "the name-bearing type is considered to be the specimen(s) illustrated or described" (ICZN 2018), not the image itself.[1] Rules in the code allow for the application of a new type when one is lost—called a neotype. However, as the hunting and capturing of certain species is outlawed, certain exceptions to this rule have been articulated. Fervent discussion has ensued regarding whether or not high-resolution photographs are a proper stand-in for physical specimens (Garraffoni and Freitas 2017; Rogers et al. 2017). One fear is that such an approach will potentially increase typification without preserved material (Krell and Marshall 2017), proving counterproductive to the purposes of types in nomenclatures rules. Another is that this creates an "inherent data deficiency problem: images cannot contain all morphological data or any genetic data possessed by actual specimens and therefore have limited utility in the face of growing knowledge" (Rogers et al. 2017, 455). As it stands, the practice is highly discouraged, at least until the scientific community agrees on a set standard, though it does seem that the issue will necessarily be reckoned with as the rate of extinction increases for species across the planet.

These type specimens stand at the center of sound and verified taxonomic work, playing a major role in correctly (as in, according to stated rules) assigning names to taxa. This is one reason museums take their conservation very seriously (figure 4.1). In a rather traditional sense, types are reference objects used by scientific professionals to initially describe taxon concepts in publications, which are required to formally instantiate a nomenclature act. As Timothy Utteridge, head of identification and naming and senior research leader at Royal Botanic Gardens, Kew, explained, subsequent taxonomists then use type specimens to confirm the connections among the publication, the type, and the actual population in the field when appropriate and necessary, especially when specimens are limited in the wild (interview 2016). The original taxon concept circumscription is,

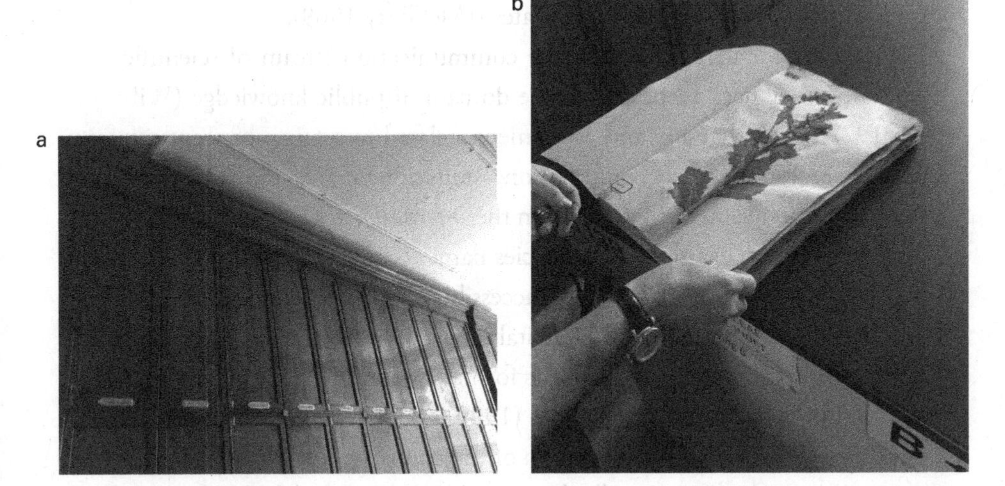

Figure 4.1
(Left) Original East India Company type specimen cabinets. (Right) A type specimen folder from the East India Company cabinet. Royal Botanic Gardens, Kew. Photo by author.

by default, connected to the type specimen insofar as it remains tethered to the name that represents the concept—any emendations to a concept must negotiate this material ground and the name it represents.

Publication and Naming

Type specimens bear the name and control the inflation of nomenclature, but the publication itself instantiates the name as a valid token to be used in scientific discourse. Publications describing a new taxon concept provide many qualitative data points of interest, including the circumscription of the species taxon, known synonymic nomenclatural variants, and type material identified, as well as a host of other potential descriptive information points (Winston 1999, pt. 3). Information contained in the publication is of central importance in the conceptualization of a new taxon, providing the necessary information for future scientists to confirm or refute a particular concept. Careful attention is paid to the characteristics that differentiate a new taxon from those that are closely related, a process known as diagnosis. Diagnosis identifies how the shared traits for taxa might differ and how taxa "differ completely" (Winston 1999, 190). In the example cited by Winston (1999, 117), the taxon concept for *Batillipes gilmartini* is differentiated from associated taxa by a "distinctly different caudal spine and the phylogenetically significant dorsal plates" (McGinty 1969).

In order for names to enter the communication stream of scientific discourse, they need to be within the domain of public knowledge (Wilson 1977), such that any and all elements related to a taxon concept can be accessed for reference, confirmation, refutation, and revision. Not just any publication venue will do. Within the *International Code for Zoological Nomenclature*, for example, for a species name to qualify as valid, publications must be permanent and freely accessible (ICZN 1999, sec. 8.1). The introduction and proliferation of digital media has required the caveat that a publication is inclusive of electronic formats so long as the content of the publication is fixed in some manner (1999, sec. 8.1.3). Good taxonomic work rests on Patrick Wilson's notion of the complete library, a documentary repository that contains all relevant published material for biodiversity work (Wilson 1977, 87).

The date of a particular published circumscription is crucial, as this imprinted date on the document is used to assess the priority of any given name over another—a critical aspect of nomenclatural standardization (Bowker 2008, 159). The identification of a publication date, however, is not always as straightforward as it may seem—a fact perhaps immediately understood by professional cataloguers and bibliographers. In conversation with zoologists at the Smithsonian National Museum of Natural History (NMNH), we learn that finding an accurate date for older publications is often difficult. Many of the zoological journals held in the NMNH dating from the early twentieth century, for example, have an accession date stamp (the date a journal was added to the NMNH collection), but do not have an imprinted published date on the issue—either by design or by a missing copyright page. Sleuthing, then, becomes a paramount task for nomenclature specialists assessing name priority. Interpolating dates from within the historical record is not uncommon, especially before the implementation of codes that required publication. Timothy Utteridge describes one such example:

> The genus I work on, Maesa that was recognized by, I think, a Danish guy . . . [on] an expedition into Saudi Arabia. And then at the same time, the Forsters went around, I think with Cook, and came back with another [sample] and they called it Baeobotrys. [The Fosters then] published this [description] and it's exactly the same time as Maesa. But the only way [we know] which one takes priority is that someone worked out when their ship landed in Portsmouth, how long it would take them to dock, how long the post carriage took from Portsmouth to London, [and] how long the editor would have taken to write it up. So, they've [assessed], to the day, which one has priority and it came out [that it was Maesa]. (Interview 2016)

These publications provide the documentary (or bibliographical) warrant necessary to prioritize one name string over a constellation of other possible name strings. The function of a warrant-based system, as Claire Beghtol (1986, 110) makes clear, is to provide taxonomists a way to understand the evolution of name types—and their surrounding publications—over a broad period of time.

INSTANTIATIVE POWER

> Naming information is the term I use for creating document surrogates. . . . I choose the word "naming" because it connotes the power of controlling subject representation and, therefore, access. . . . Theories, models, and descriptions are elaborated names. In these acts of naming, the scientist simultaneously constructs and contains nature.
>
> —HOPE A. OLSON
>
> *The Power to Name: Locating the Limits of Subject Representation in Libraries* (2002)

Biodiversity and IS work both are heavily involved in the production of names and their subsequent control—names that ultimately impact the lived natural and social worlds. I think it is relatively easy to get lost in the procedural aspects of nomenclature production and taxon concept construction, particularly when the process is so codified by way of rules of nomenclature, delineation, and description. The impact of this process, though, should not be underestimated, as it lays out the processes by which formal taxon units are created, on which all taxonomic work rests. Biological names instantiate concepts, which can then be applied and operationalized in the process of taxonomy and class construction for real world entities. To have this kind of instantiative power is to maintain control over what can and cannot be entered in the intellectual field of record, where species are formally articulated as existent or not within a host of biodiversity databases, records, and documents. In the case of our imperiled dingo, the stakes could not be higher when we consider whether a certain formal infrastructure (whether the Catalogue of Life or some other database) adopts the formal designation of *Canis lupus dingo* or *Canis familiaris dingo* as the prioritized name and the interpretation of its associated taxon concept. And certainly, the best taxon concept construction work requires that the full historical trajectory of any given concept be available to make the best determination based on the preponderance of scientific evidence.

Within the bibliographical realm, name production and instantiation have a very particular application as well, that serves as an analogue for change of a different sort. In Patrick Wilson's *Two Kinds of Power*, instantiation is understood as the production of "instances of particular patterns, or types" (1968, 7). In Wilson's context, an instance of a work, for example, is the physical production, performance, or exemplification of some abstract creative endeavor—a book, a play, a song, or the like. Work-instantiation theory, then, looks to formulate formal relations between a work and its copies (see for example, Smiraglia 2001, 2005). Textual criticism and textual bibliography (Gaskell 2007, 313–360), for example, are domains of study that examine the genealogy and editing of texts over time. Or, put another way, these areas follow the production and alteration of work-instances over time. In textual bibliography, for example, work-instance variations might be "compared" with what is called an "ideal copy." An ideal copy, as espoused by Fredson Bowers, is an "ideally perfect" copy of a text as the printer or publisher (and, I might add, the author) originally intended it (1994, 113). From this idealized version of the work, variations and errors introduced during publishing and production can be identified and documented against. G. Thomas Tanselle's (1980) version of ideal copy is a bit more materially focused and composite-based than Bowers's, given that the "reconstruction" of an ideal copy "encompasses all states of an impression or issue, whether they result from design or accident" (1980, 46). As Tanselle continues, "the '*ideal copy*' is central to descriptive bibliography, because it is the element that distinguishes bibliographic description from cataloguing: whereas a catalogue entry, regardless of its level of detail, exists to record a particular copy, a bibliographic description [centrally using the concept of the ideal copy] aims to provide a standard against which individual copies can be measured" (1980, 21; emphasis original). The ideality presupposes a printer's intended perfect instance of a text, which means that, in practice, these ideal copies are more often than not abstract, as this "ideality" is not attainable.

Much like taxon concepts, works and texts are also constantly shifting and evolving along temporal and spatial lines—albeit on very different terms. The identity conditions of a work change over time depending on how its work-instances persist (materially), are reformulated (republished or edited), emended (critical editions, adaptations, and the like), or reinterpreted. Much like Wilson's notion of the subject, the line between one work and another is infinitesimally graded, requiring the articulation of artificial boundaries to differentiate them. Within information systems, such as a library catalogue for example, the challenge is to draw lines between various instances of works so that the significant differences between each entity can aid in the selection of appropriate resources. In some ways, however, the decision-making process for bibliographers and cataloguers is easy compared to that of the taxonomist studying the minute differentiation of concept change.

On Fish and Control

Even with the concept of nomenclatural priority in effect, tracing the historical records associated with a name is an onerous task. Imagine if name inflation wasn't controlled, how much more difficulty would be encountered as the name disambiguation process played out? Inevitably, if one is to understand the full context of any given name and associated taxon concept, one must take at face value that the circumscription and delineation of the taxon is as fully represented and articulated in a given publication as possible, and that the appropriate references and taxonomic determinations have been documented. And a crucial part of making taxon concept determinations is being able to represent and usefully follow the nuanced changes for that proposed taxon group over a broad period of time. Complicating matters, taxon concepts are anything but stable, particularly in some taxonomic groups. As noted by Nico Franz (2010), only approximately 55 percent of the valid concepts in "eight succeeding classifications of North American vascular plants from 1933–2006" remained stable (2010, 49). The diagnosis section for a new species taxon circumscription, for example, requires taxonomists to examine publications and samples for all closely related species, such that distinctions can merit its inclusion as a new concept. Similarly, revising a taxon requires that one pay special attention to how a particular taxon has permutated over time. The publication of a new or revised species taxon most often also requires a taxonomic discussion section, which outlines the logical reasoning of the taxonomic assessment (Winston 1999, chap. 12). Careful examination of literature (often mostly accessible through nomenclature tokens in a database) is necessary to properly and ethically articulate new grounds for a taxonomic revision. Quite simply, if taxon concepts-as-names are not readily available, and mapped, the practice of taxonomy comes to a virtual standstill—or worse yet, prior mistakes or observations risk being duplicated or ignored. Preserving the historical continuity of taxonomic knowledge is paramount. As we will see, this kind of change is not usually defined by a simple, linear process. To illustrate how complex change becomes in the practical realm, let us look to an example by Richard Pyle (2008), senior curator of ichthyology at the Hawaii Biological Survey of the Bishop Museum.

Imagine two hypothetical species of fish are extracted from a pool of water in the wild believed to be part of the same genus, named as:

Fish 1: *Holocanthus fisheri* (Snyder 1904) [sec.] Snyder 1904[2]
Fish 2: *Holocanthus acanthops* (Norman 1922) [sec.] Norman 1922

Then Jordan comes along and decides that *Holocanthus fisheri* is actually part of another genus, *Xiphypops*, so he renames it with a new combination moving the genus:

Fish 1: *Xiphypops fisheri* (Snyder 1904) [sec.] Jordan 1922
= *Holocanthus fisheri* (Snyder 1904) [sec.] Snyder 1904

But notice that the concept hasn't changed; it has the exact same circumscription (description) as Snyder 1904.

Then imagine a third scientist described another fish from that same pool, and describes the following as part of a new genus:

Fish 3: *Centropyge flavicauda* (Fraser-Brunner 1933) [sec.] Fraser-Brunner 1933

But Fraser-Brunner also thinks that *all of the fish* from this pool are from this same *new* genus, so she decides to move *all* of the others into the same genus as well:

Fish 1: *Centropyge fisheri* (Snyder 1904) [sec.] Fraser-Brunner 1933
= *Xiphypops fisheri* (Snyder 1904) [sec.] Jordan 1922
= *Holocanthus fisheri* (Snyder 1904) [sec.] Snyder 1904

Fish 2: *Centropyge acanthops* (Norman 1922) [sec.] Fraser-Brunner 1933
= *Holocanthus acanthops* (Norman 1922) [sec.] Norman 1922

In his presentation, Pyle continues to describe how yet another individual comes along and decides that Fish 3 is actually a synonym of Fish 1, and proceeds to bring those two species groups under one genus:

Fish 1 and Fish 3: *Centropyge fisheri* (Snyder 1904) [sec.] Pyle 2003
> Fish 1: *Holocanthus fisheri* (Snyder 1904) [sec.] Snyder 1904
> Fish 1: *Xiphypops fisheri* (Snyder 1904) [sec.] Jordan 1922
> Fish 1: *Centropyge fisheri* (Snyder 1904) [sec.] Fraser-Brunner 1933
> Fish 3: *Centropyge flavicauda* (Fraser-Brunner 1933) sec Fraser-Brunner 1933
= Fish 1: *Centropyge fisheri* (Snyder 1904) [sec.] Fraser-Brunner 1933
+Fish 3: *Centropyge flavicauda* (Fraser-Brunner 1933) [sec.] Fraser-Brunner 1933[3]

It was a circumstance such as this that led Pyle to remark jokingly during the 2008 annual Biodiversity Information Standards/Taxonomic Databases Working Group (TDWG) meeting, "Taxonomy is the perpetual classification of mis-named species" (2008). Multiple names can represent the same taxon concept (synonyms); one name can be used for many entirely different concepts (homonyms); and one name can refer to two or more concepts, whose circumscriptions overlap (usually resulting when taxa are split or merged over time) (Remsen 2016).

Herein lies the problem, according to Pyle: "Sometimes the same concept goes by different *legitimate* names and sometimes the same name can refer to different *legitimate* concepts" (2008; emphasis added). In light of this, the historical network of taxon name tokens becomes increasingly difficult to parse and differentiate. As Pyle's example shows, name formulations have the capacity to include semantically embedded metadata (author, date, and so on). But such information is, at best, only occasionally included as part of a name string. Most name-forms do not include these provenance markers, despite calls from taxonomists that they be consistently included (Franz, Peet, and Weakley 2008). Nico Franz has also been vocal about the importance of computational ontologies to bring order to nomenclatural issues such as those described above (2010). And although these ontologies cannot solve all the problems inherent in nomenclatural control, they certainly have more ability to map the qualitative differences among name tokens in ways more informative than hierarchical presentations or traditional nomenclatures. Relationship indicators such as "synonym of" can produce more meaningful relationships that also have the ability to be mapped over time (Groß, Pruski, and Rahm 2016).

Concept determination and concept change play out in different ways in IS, of course, but concept mapping is equally important in both fields. One analogue is how we go about determining the historical evolution of subject terms in classification systems, and how these studies can illustrate certain cultural and academic trends over time. As Joseph Tennis (2002) relates in reference to his subject ontogeny studies, "The power of a classification scheme to collocate is compromised if we do not account for scheme change" (2012, 1350). Scheme change, like taxonomic change, is useful not only as a document of the classification itself, but also to facilitate information retrieval as subject and access terms evolve—a matter not wholly different from the problems faced with nomenclature in the biodiversity taxonomic world. As noted by Ellen Greenblatt, subject term change can represent the mapping of similar terms in syndetic relationships, such as mapping the terms "lesbian, dykes, and gay womyn" (2010, 212). And, akin to Tennis's study, Greenblatt also notes that obsolete terms must also be mapped onto their contemporary permutations, such as how the "terms *lesbigay* and the more inclusive *lesbigaytr*," once in vogue in the 1990s, have now fallen out of colloquial use (2010, 212). Certainly, as more robust infrastructure increasingly arises to more efficiently represent taxonomic changes over time, IS should be poised to take lessons from taxonomic professionals that deal with such change on an exceedingly more complex level.

CONTROL, OUTLINED

The first step toward nomenclatural control is aggregating all undifferentiated name-forms produced around the world, ranging from those that are in correct scientific form to vernacular and common names to any possible iteration between these two poles. In this initial stage, names have not yet been disambiguated or validated by nomenclature professionals. Valid names are both well-formed (syntactically) and attend to the nomenclatural rules specified within their particular domain in terms of both syntax and semantics (independent rules exist for the botanical, zoological, and viral taxa, for example).

The Global Names Architecture (GNA) has arisen to serve this vital name-collocating function in the biodiversity world (GNA 2021c). The GNA locates and indexes name instances scraped from across the web and links these name instances directly to their original source (2021b). The GNA is composed of two distinct parts: the Global Names Index (GNI) and the Global Names Usage Bank (GNUB). The first and less curated of the two, the GNI, is a list in the broadest sense, and includes any number of name-forms, including code-compliant scientific names, common names, genetic barcodes, and any other generic species identifiers (GNA 2020a). Any token that points to a species or species concept is included. Because of its catch-all nature, the space is often jokingly referred to as the "dirty bucket" (GNA 2020a). As of April 2021, the GNI data bank contained more than seventeen million name-forms (GNA 2021a). Content for the GNI is variable, unstandardized, and contributed by many organizations and sourced from both online and digitized analog sources, including articles, databases, websites, and online specimen repositories where labels and other museum data are scanned with OCR software for easy ingest. Institutions such the Natural History Museum, London; Royal Botanical Gardens, Kew; and the Smithsonian National Museum of Natural History regularly contribute digitized literature. Given this distributed harvesting methodology, the GNI necessarily contends with hundreds, perhaps thousands, of list sources containing thousands or hundreds of thousands of names, many of which are far less curated than those at large, prosperous institutions such as Kew. Error reconciliation and name disambiguation become the greatest challenge in this environment—a matter we will deal with in just one moment.

Associations

It benefits our narrative to mention that names within the GNI do not just float without reference; they are tied to their documentary context. Associations between a name and a document source become especially important once nomenclature specialists begin validating taxon concepts and assessing the priority of token forms in the pool of potential concepts. But the relationship between a text string in the GNI and its documentary source is relatively flat and provides very little metadata related to the name's full relationship within its source. A mere one-to-one linked relationship makes it difficult for algorithms to link related and alternate name forms together in useful concept clusters (Pyle 2016). More robust relationship-building is then necessary.

The GNA's Global Name Usage Bank is just such a mechanism. GNUB takes each individual name token reference and associates it with its full, institutional context. As explained by Pyle, the GNUB provides a persistent globally unique identifier (GUID) to all potential entities involved with the names instance, which includes agents (people and organizations that are responsible for the usage), as well as references, including published literature as well as unpublished reports, manuscripts, specimen labels, herbarium sheets, field notes, and the like, used in a taxon concept's circumscription (2016, 270–271). Within this network, any one reference document can contain many taxon name usages (TNU), just as any author can be associated with many published documents. With these associations in place, the disambiguation of one name-form from another can now proceed, which, given the size of the GNA, can be completed only through computational mediation.

Resolution

Linking harvested names to their sources, authors, and instances is just the first of many data transformations necessary to make this nomenclature useful. The next step, which emphasizes the syntactical elements of names, requires that scientific names be identified, extracted, and validated from all other nonstandardized name tokens. Whether or not colloquial or common names are integrated into the system must be decided at this point. Although the Catalogue of Life prioritizes code-compliant scientific names, common names are also widely available in its listing (see box 4.1). In a

Box 4.1

The Catalogue of Life has defined **fourteen field groups to be the standard set of data** for each species (or infraspecific taxa).

1. **Accepted Scientific Name** linked to **Reference(s)** (obligatory)
2. **Synonym(s)** linked to **Reference(s)** (obligatory, where available)
3. **Common Name(s)** linked to **Reference(s)** (obligatory, where available)
4. **Classification above genus, and up to the highest taxon in the database** (obligatory, where available)
5. **Distribution** (obligatory, where available)
6. **Life zone** (obligatory, where available)
7. **Current and Past Existence** (obligatory, where available)
8. **Additional Data** (optional)
9. **Latest taxonomic scrutiny** (obligatory)
10. **Reference(s)** (obligatory, where available)
11. **Taxon Globally Unique Identifier** (obligatory, where available)
12. **Name Globally Unique Identifier** (obligatory, where available)
13. **Catalogue of Life LSID** (obligatory)
14. **Source Database** (obligatory)

Catalogue of Life Standard Dataset Field Groups. Species 2000 has defined fourteen field groups to be the standard set of data (version 7, September 23, 2014) for each species and infraspecific taxon in the Catalogue of Life. (Species 2000 2016c). Used by permission.

perfect world, each name string is handled by human hands, though such attention is not possible in a pool of seventeen-million-plus names. Name matching services based on various lexical algorithmic software become crucial to facilitating this process (Pilsk, Kalfatovic, and Richard 2016; Vanden Berghe et al. 2015). The GNA has its own Global Names Resolver (GNA 2021b) that examines text strings to assess whether a name is in scientific form, correctly spelled, and currently in use, along with a host of other metrics (2020a). But algorithms and name resolvers have their limitations and can introduce any number of data failures and aberrations.

Scientific names, for example, must be Latinized and are most often binomial, consisting of two parts: a generic epitaph (genus name) and a specific epitaph (species name) (though variations to this rule do exist); *Ursus arctos* is an example of a correctly formatted scientific name. But as the GNA documentation points out, problems quickly arise during this process, since not all binomial Latinized name forms refer to species, such as with the terms anorexia nervosa and habeas corpus (GNA 2020b)—neither of which point to species, though either could be chosen by an algorithm as satisfying the syntactic standards for inclusion into a formal listing. Additionally, in some cases, orthographic conventions between the botanical and zoological codes sometimes call for different (and conflicting) name-forms that just cannot be resolved by automated means. Such shortcomings of GNUB are too significant for organizations such as the Catalogue of Life, and so alternate control mechanisms have been created to mediate these possible introduced errors.

The semantic value of names must next be addressed. Nomenclators identify the valid and accepted names that can be used for subsequent taxonomic work. But tracing the history and development of valid scientific names, and then subsequently disambiguating taxon concepts from one another is necessarily very human, time-intensive work. Priority requires understanding publication dates. To decipher one taxon concept over another, for example, nomenclators must examine all associated taxon documentation, including potentially the publication (with the detailed circumscription), as well as the type specimen. Nomenclator listings, then, are expected to be finely curated by way of expert review by assuring name currency and consistency and associating valid names with their associated publication dates (Croft et al. 1999, 320). Like thesauri, nomenclators are expected to establish primary taxon names and their synonyms. The result of this process is a controlled list of names that constitute code-governed facts. An example of a robust nomenclator is our previously discussed International Plant Names Index, located at the Royal Botanical Gardens, Kew. What emerges are clusters of related terms that contain a central, validated current name-form, as well as its associated synonyms and homonyms, similar to Pyle's final fish population, reproduced here:

Fish 1 and Fish 3: *Centropyge fisheri* (Snyder 1904) [sec.] Pyle 2003
> Fish 1: *Holocanthus fisheri* (Snyder 1904) [sec.] Snyder 1904
> Fish 1: *Xiphypops fisheri* (Snyder 1904) [sec.] Jordan 1922
> Fish 1: *Centropyge fisheri* (Snyder 1904) [sec.] Fraser-Brunner 1933
> Fish 3: *Centropyge flavicauda* (Fraser-Brunner 1933) [sec.] Fraser-Brunner 1933
= Fish 1: *Centropyge fisheri* (Snyder 1904) [sec.] Fraser-Brunner 1933
+Fish 3: *Centropyge flavicauda* (Fraser-Brunner 1933) [sec.] Fraser-Brunner 1933

Mapping out these name transformations over time makes for an exploitable information system in that users can assess, to a high degree of accuracy, the best species concept for their purposes. This synonymic amplification is a vital feature of effective biodiversity information seeking (Guala 2016).

THE CATALOGUE OF LIFE PLUS

To those designing the Catalogue of Life, it was clear early on that its hypercurated nomenclatural environment could be a hindrance to its widespread adoption and use. One can present many circumstances in which a perfectly valid taxon concept might not be present in the Catalogue. Recent publications describing new taxon concepts, for example, would not be represented in the system, given its update cycle. Further, there might not be enough human labor or expertise involved to meet increasing demand for more comprehensive lists. Still more onerous is to produce a list that includes a widely expanding groups of name variants, including a multitude of common names. The Catalogue of Life Plus is attempting to meet this challenge. This initiative is a partnership between Species 2000 and GBIF that resulted from the second gathering of the Alliance for Biodiversity Knowledge meeting (GBIF 2019). The CoL+ is meant to bridge an undifferentiated GNA-type space with the highly curated space of the Catalogue (Species 2000 2020 [2017]). The goals for the Catalogue of Life Plus are to

1. create both an extended and a strictly scrutinized taxonomic catalogue to replace the current GBIF Backbone Taxonomy and Catalogue of Life;

2. separate nomenclature (facts) and taxonomy (opinion) with different identifiers and authorities for names and taxa for better reuse;
3. provide (infrastructural) support to the completion and strengthening of taxonomic and nomenclature content authorities;
4. ensure a sustainable, robust, and more dynamic IT infrastructure for maintaining the Catalogue of Life.

The CoL+ is thus a layered clearinghouse system, whereby each layer represents a different level of curatorial control (see figure 4.2). The operational taxonomic units (OTU) in the outer cloud of figure 4.2 represent undifferentiated species concept tokens—a scientific name, a common name, or a genetic barcode. Users throughout the globe can then theoretically (given future infrastructural enhancements) propose relationships between OTUs in the outer cloud with scientific name-forms held in the middle cloud, here titled, "Col Plus/Linnean names." For example, an expert user in the genus *Abies* (a group including fir trees), could propose links between the

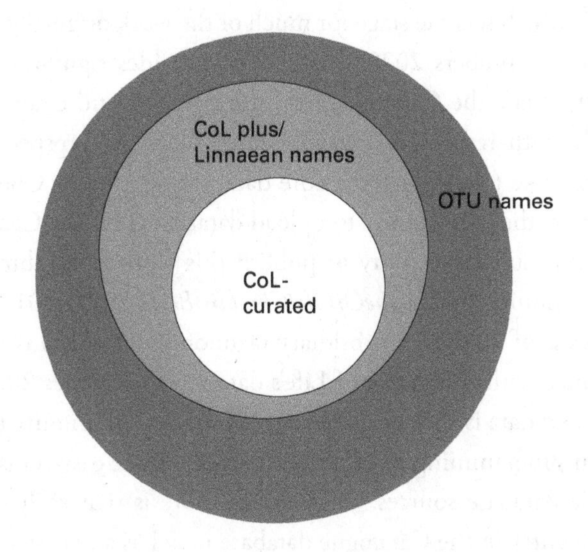

Figure 4.2
Catalogue of Life Plus Layer general schematic. Central names are curated to a gold standard. Linnaean names are names that are semantically meaningful and ready for ingest by the central repository. The outer ring consists of semantically incongruent forms that must be reconciled by volunteers before being ingested into the workflow.

common name "balsam fir" with its genetic barcode. Even more enticing, the CoL+ also grants the possibility for some a limited epistemic expansion of the system. What if, for example, individuals knowledgeable of indigenous botanical associations could also associate *Albies* with the Cherokee name for the same, *a'ninandak'*? Of course, this does not sidestep the general problematics of Western biological naming tradition (as will be discussed in future chapters), but it certainly does allow for more radical and diverse cultural enhancements of biodiversity knowledge. All of these associations can then be tentatively associated with the verified, scientific name *Abies balsamea* (L.) Mill in the CoL+ database layer. Such suggestions could then be flagged for validation by Catalogue editors and, after evaluation, allowed entry into the Catalogue's validated-name space. The result is a more nuanced, crowdsourced database that offer infinite opportunities to expand the horizons of nomenclature spaces.

In December 2020, the Catalogue of Life, in conjunction with GBIF, published a rebuilt infrastructure for what is now termed the Catalogue of Life Portal, which sets the stage for much of the work originally envisioned in the CoL+ (Huijbers 2020). The portal provides significant enhancements that situate the Catalogue for future growth and citation stability, particularly with respect to invalid names, with the prospect of slowly integrating these tokens into the core database set. A new ChecklistBank provides users the opportunity to upload databases into the Catalogue system, and thus an opportunity to publish this work to the broader taxonomic community (*CatalogueOfLife/ChecklistBank* 2021 [2017]). As part of the process of uploading subsidiary taxonomies, the data is normalized in accordance with Catalogue of Life's data model (Species 2000 2021a). All aggregated data is then available to the broader community through an application programming interface (API) to facilitate easy transfer of data to multiple database sources. Of technical note is that each species and higher taxa entry in the Catalogue database now has a persistent identifier that connects the current and archived data sets. With persistent identifiers in place, users and subsidiary databases can integrate the Catalogue and seamlessly update their repositories with each annual release without any identification conflicts. This new portal and ChecklistBank sets the stage for a significant broadening of the core Catalogue nomenclatural set and taxonomic core.

The importance of developing systems such as the Catalogue of Life Plus cannot be underestimated, particularly because it is through these mechanisms that more inclusivity is built into the process of instantiative control. Concept control begins with the basic process of identifying the entities that constitute our field of concern, which, herein, I've operationalized as a process that involves producing species taxon concepts by way of the formal integration of evidence, producing nomenclatural tokens that are then appended to particular taxon circumscriptions, and subsequently mapping these name and taxon concept associations as scientific opinion evolves over a broad period of time.

The reality, of course, is that names make taxon concepts into, as Ron Day calls them, meaningful things, within a larger network of biodiversity practice (2014, 6). Names are tangible and machine-readable and, as tokens, are what systems are designed to manage and collocate (Furner 2016, 120), even if, as we will find, they fall short in many respects as monikers for complicated concept histories. I cannot stress the importance of this fact enough: our ability to accurately classify and represent the world (natural or otherwise) is directly proportional to the control of language terms that we have at our disposal to represent it. In chapter 5, we turn our attention to classifications-as-systems, including how they function as mechanisms of reduction, universality, and, ultimately, spaces of epistemic authority. As we have seen, the Catalogue is not only a space for nomenclatural control, it is also a taxonomic management system. It is here, at the level of the taxonomic whole, that we begin to see the structural expression of power take shape and the vast implications this has on scientific activity and the human imagination, broadly construed.

5 EPISTEMIC POWERS

> Every proposition proposing a fact must, in its complete analysis, propose the general character of the universe required for that fact. There are no self-sustained facts, floating in nonentity.
>
> —ALFRED WHITEHEAD
> *Process and Reality* (1985, 11)

TAXONOMIC STRUCTURE AND INTERNAL INTEGRITY

Power is a structural concept, making power both pervasive and elusive. In Miranda Fricker's (2009, 4) *Epistemic Injustice: Power and the Ethics of Knowing*, she defines social power as the "socially situated capacity to control other's actions." As human-made artifacts, all classifications are, at their core, expressions of social power. That power, as we have seen, can be epistemic or material, beneficial or detrimental. The primary role of biodiversity taxonomists is to posit empirically informed, hypothesis-based arrangements of living entities in natural world. They take expressed nomenclature and its associated taxon concept and propose unified taxonomic systems comprising species and their relationships. This is, at first glance, functionally different from the purpose of bibliographical or documentary systems, which are touted more as a means to facilitate retrieval than as a representation of the universe of knowledge *as it should be.* Indeed, bibliographical classifications, unlike those in biodiversity, are pragmatic, based more on the social organization of knowledge production (Bliss 1929). However, even if unintended, these practical bibliographical classifications have derivative impacts on how we value some forms of knowledge over others—and this has a great deal of epistemic influence. All classifications help us make sense of a complex world. Classification systems like Google's search engine, for example, take an incomprehensible volume of information and deliver it to us in ways that are digestible and, ideally, useful. By participating with this system, however, we begin to see and interpret our online world through the narrow and opaque lenses of algorithmic mediation.

In this way, classifications have an instrumental function, in line with Patrick Wilson's discussion of "bibliographical instruments" (Wilson 1968, chap. 4). Wilson believed that, to understand how much control we have within a bibliographical system, we must understand how it is logically arranged. Wilson states, "I cannot tell how much bibliographical control I have or could have simply by introspection, by memory of past success and failures, or by flexing my muscles. To discover what I can or might do if I would, I must discover what arrangements there are of which I can take advantage, what bibliographical instruments . . . are at my disposal" (Wilson 1968, 55–57). We must understand, for example, what it means that a subject inhabits one position in the instrument over some other position. And by extension, the better we understand an instrument's structure, the more we can understand the inferences that can be drawn from it. This chapter, then, focuses on these systemic aspects of classifications. I further argue that a classification's structure necessarily forms reduced and univocal interpretations of the world. I discuss classifications as exerting structural power in three senses: as mechanisms that (1) express which entities do and do not merit annunciation in the world; (2) obfuscate the inherent complexity present in nature; and (3) present nature in a univocal fashion, at the expense of alternative forms of knowledge.

Let us imagine a situation in which two taxonomists are given a "pile" of flash cards that represented a suite of taxon concepts and are asked to classify them in accordance with their own taxonomic commitments. Their first operation will be to take these taxon concepts, group them into species groups, and articulate relationships among them according to their preferred theoretical approaches to classification. For example, one might relate species by way of phylogenetics, whereas the other might choose to use phenetic approaches. Let us also further assume that we ask both these scientists to employ the same approach to constructing relationships—for example, doing so based on overall similarity using morphological approaches in accordance with phenetics. The chances that the two resulting taxonomies will be exactly the same are incredibly small, if nil, even if they both approach classification using the same theoretical foundations. One may lump more than split, thus creating a taxonomy with fewer overall species, while the other may mark small morphological differences as reason enough to merit a new species taxon, resulting in many species branches. This diversity of opinion is a natural part of the taxonomic landscape and the reason such taxonomies are said to be hypotheses. As stated by Clare Beghtol, "every classification system is a theoretical construct imposed on 'reality'" (2001, 99). And reality, as we know, is relative.

One rule of biological taxonomies (traditional ones, in any case) is that they must maintain a certain degree of internal integrity and consistency with regard to how they represent the external world. Taxonomies are only credible, after all, to the extent that their classifications maintain their commitments throughout their structure, in accordance with user's assumptions about how reality is composed. That the Catalogue of Life does not display this internal coherence is a prime criticism of its composition. As identified by Furner (2009b, 9), how we judge the success of a knowledge system has been a question of great concern to those seeking to classify knowledge, particularly in relation to how effectively these systems represent the external world. Taxonomies should be coherent, in that the rules of arrangement should be invoked uniformly within the classification. When a given taxonomy is consistent, a user can understand the logic behind how and why a given species is at a given location. Some believe that another hallmark of good classifications is that they should be simple and elegant to the extent that their taxonomic hierarchies are parsimonious and describe "the evolution of any particular set of characters using the smallest number of evolutionary changes" (Wiley and Lieberman 2011, chap. 6).

Yet we know that taxon concepts and taxonomic classifications change and are challenged over time. Taxonomic knowledge systems are products of contextually specific historical, cultural, and philosophical circumstances that evolve. Understanding that all classifications are unstable and subjective, even and especially within the natural sciences, helps me illustrate one of the general limits of universal classification systems: they are never truly correct or final. Universal systems, however constructed, are never truly universal. Further, the subjects we embed within them are not real, or definite, yet we continually treat them as if they are. And this truth applies to species: species taxa are not definite concepts, yet people outside of the biodiversity sciences continually speak about them as if they are statements of fact. As Richard Rorty wrote, "The desire for a theory of knowledge is a desire for constraint—a desire to find 'foundations' to which one might cling, frameworks beyond which one must not stray, objects which impose themselves, representations which cannot be gainsaid." (2009, 315). Giving structure to the abstract is one of the primary purposes of any classification—a theory for the presumed order of knowledge of natural biodiversity—biological or otherwise. Nature is, ultimately, abstract, and we necessarily create linguistic metaphors to make sense of it.

CONTINUITY AND CONTINGENT EXISTENCE

In light of this abstraction, let us now switch our focus to the backdrop, so to speak, of biological classification. I want to expand on a rather large and amorphous concept: *nature*, and why I think it is important for me to assert, at least when discussing classification matters, that nature does not have natural joints, and that classifications have the often-negative popular effect of making it seem as if they do. If the experience of doing fieldwork with biodiversity scientists taught me anything, it was that saying as much to a room full of highly educated and accomplished individuals set the groundwork for some fiery (but lighthearted) repartee. And this is understandable; to say that there are no true, individual entities in nature and that, if we say there are, they are social and not natural, might sound like a slight against the informed, elegant work of taxonomists. Surely their work is incontrovertibly valid, and increasingly more essential, to the study of our natural ecosystems, especially given the challenges we face with a changing climate and rampant extinction. So, let me first establish what I do not mean. I do not mean that there are not, in fact, distinguishable natural entities in the world that are potentially categorizable as distinct. My argument is not an ontological one, questioning the existence of perceptively discrete concrete objects as possible entities that exist in the natural world. To my mind, there are organisms that scientists try to describe in this world—animals, plants, bacteria, and the like—and whether I am a materialist or an idealist makes no difference to my central argument, which is that nature does not produce simple classifications of organisms for us. Surely, when I was camping a few years back, when a pair of grizzly bear cubs rose to their hind limbs in front of me, the first question that came to mind was not, Are they (and their likely nearby parents) real? A grizzly is surely concrete enough that I'll back away slowly regardless of my ontological inclination. My assertion, of course, is much more nuanced than that. My argument is not that biological classification is wholly arbitrary, but rather quite the opposite: to be a taxonomist is to be carefully trained in method and to make informed decisions about what elements in nature *matter most* to distinguishing one living organism from another.

Species taxon concepts, however, *are* contested. As Kriti Sharma states, "No one expects a flower to be unitary—we know that some people will define the flower as a bloom atop a stem, and others will define it is as the whole plant down to its finest roots" (2015, 13). Nature, from the point of view of classification work, is most usefully defined as a process, and if there are any joints to be found, they are merely temporary. At the heart of this argument is, to a certain extent, an organicist approach to biological entities—a process-oriented philosophy that believes process, not things, to be, if not the most ontologically fundamental category, then certainly one of the primary characteristics of the natural world (Nicholson and Dupré 2018, 12–13). The view of process thinking is that life can be viewed as one process, or rather as a series of processes that mutually constitute one another at varying rates of transition. "A process-based metaphysics," Argyris Arnellos writes, "implies that living systems (such as cells, multicellular systems, ecosystems, organisms) are just temporary phenomena—the products of the dynamics of some processes" (2018, 200). Nature, as it were, is a set of processes, interconnections, and moments of transfer. Alfred Whitehead (1920, 13, 65) calls this "complex of nature" an undifferentiated fact that becomes a "entity for thought" that is defined, in part, by the innumerable interconnections between a string of "extensionless instants" perceived by a given viewer. Whitehead's philosophy presumes that objects, insofar as they are perceived, are *events* that are successively realized over time (1920, 126). Objects-as-events, then, are historically or genealogically connected perceptions that are effectively different each time they are experienced. The plant we see in our living room one day is temporally distinct, and thus a different event-object than the one previously perceived.

In *Interdependence: Biology and Beyond* (2015), Kriti Sharma argues for a systemic model of nature that highlights a dynamic, mutually constituted notion of the natural-world-as-entity within the biological sciences. Sharma describes two shifts that define this line of thought:

> The first is a shift from considering things in isolation to considering things in interaction. This is an important and nontrivial move. . . . To get to a thoroughgoing view of interdependence, I argue that a second shift is required: one from considering things in interaction to considering things as *mutually constituted*, that is, viewing things as existing at all only due to their dependence on other things. (2015, 2; emphasis original)

Sharma's approach is enticing, particularly because they make clear our perception itself is entangled in this mutual constitution and it is our perception that necessarily creates classes to make sense of a radically dependent space. This is not to say, as Sharma notes, that objects cease to exist in the world if we do not see them—after all, the grizzly bear that chases me away does not fail to be a grizzly bear (nor to be dangerous!) once I (or all people) flee to a safe distance. But, at the same time, without me, or anyone else as an observer, there is indeed nobody there to actively characterize that bear as being bear-like, brown, and in possession of skin-piercing, potentially murderous claws. "Only observers can perform the various actions necessary for experiencing phenomena as objects," states Sharma (2015, 22). We adhere attributes onto objects through naming and language. Objects require subjects to instantiate them within an ontology of other objects and relations.[1] To be the object, *tree*, for example, means that the object must fit with our notions about what we believe to be a tree and be defined in contradistinction to a host of other objects that do not express the tree-like properties of a tree (bushes, weeds, grasses, and the like). One result of this analytic approach is that, while objects of "nature" can exist independent of the subjects that encounter them, once we sense them, classify them, and name them, they become socialized and articulated within our instruments of knowledge.

The problematics that such a processual view might have on the definition of species are quite obvious: if, indeed, everything is mutually contingent and event-based, how do we get to the point where concrete objects of any kind are distinct and locatable in time and space? In thinking about the limits of the body, Whitehead stated, "It is just as much part of nature as anything else there—a river, or a mountain, or a cloud. Also, if we are to be fussily exact, we cannot define where a body begins and where external nature ends" (1938, 21). A conundrum, indeed. John Dupré (Dupré 1993, 2014; Nicholson and Dupré 2018) has been a leading thinker in this area. "Many processes are bona fide individuals—they are concrete, countable, and persistent units. . . . In biology, processes are . . . dynamically stabilized at vastly different timescales: a matter of minutes for a messenger RNA molecule, a few months for a red blood cell, many decades for a human being, and up to several millennia for a giant sequoia tree" (Nicholson and Dupré 2018, 12). Temporal scales are an essential component in process theory, given that processes are not uniform in their velocity or rate of change. As humans, we are able to see the full systemic life cycle of a red blood cell or an RNA molecule. But witnessing a species evolve from one to another is an entirely different matter. Evolution, for one, is a property not of an individual member of a species but of populations and, as such, works slowly over the course of time. Watching evolution happen to a population is an impossible ask of any scientist. Thus, scientists look for other forms of evidence to help measure difference between genealogical lineages (phenetically, genetically, and so on) and infer evolutionary changes. An entity can appear concrete by way of an equilibrium of many intermingling dynamic systems functioning at a speed that exceeds the perceptual abilities of the viewing subject. A distinguishable eddy in a river, for example, is one concrete object-system that flows harmoniously as part of a larger river. As is an organism that lives within a larger ecosystem among many other seemingly distinct species.

The obvious reality is that we *do* sense and experience entities, and this is not antithetical to process-theoretical accounts. Dupré advocates for what he calls "promiscuous individualism," given that "there are various ways of drawing such boundaries, reflecting real biologically salient aspects of the multiply interconnected systems that make up the biological world" (2014, 241). The approach accepts that there can be many equally valid mechanisms to divide entities from one another. Depending on one's methodological approach, some systemic interactions appear in the fore- or background of an analysis. But, make no mistake, the dividing line between one cohesive entity-system and another is, indeed, blurry. An individual, then, in Arnellos's assessment, reflects the fact that "some [systems] work together so that they constitute *cohesive systems*, that is, systems that persist and manifest in a *form of stability* in the sense of a spatiotemporal integrity" (2018, 203; emphasis original). The individual can be described as a self-enclosed, constitutive system, constrained in such a way that matter and energy can be maintained, while still remaining open enough to engage with the external systemic environment (2018, 206).

If we were to adopt Kriti Sharma's (2015, 12) terminology, we'd say that things as they exist in our conception of the world—including objects such as organisms—exhibit themselves through "contingence existence." To exist contingently is not to exist individually and inherently outside of any given process, but rather to recursively depend on other equally contingent objects and subjects for being (2015, 15). So, like Whitehead and Dupré, Sharma sees the environment as mutually constituted by a series of ongoing object-object and object-subject relationships. Nothing exists on its own inherently, and as such, when we classify and categorize, it is imperative to understand the ecological conditions under which these entities came to be. This is an imperative concept to understand within the domain of information studies, primarily because it forces us to see that our classifications (of books, of documents, of species, of people, and so on) are entirely contingent not only on the internal, intellectual context of the classifier, but also on the emergent and inherently fuzzy boundaries between entities and concepts in the external world.

Such a view forces us to examine intersections of beings, identities, and qualities in light of their subjective and relational qualities. We *should* question the application of subjects—vis-à-vis Melissa Adler and Hope Olson's work—because any subject, or any combination of subjects, we apply to a document can never comprehensively describe it adequately. In addition, a much more ecologically and biologically focused driver is that we have to begin the long, complex process of reorienting our relationship to biological objects, the natural world, and the affective impact our classifications have on the entities that are increasingly vulnerable in the world.

REDUCTION AND OUR MENTAL REALITY

Classifications extract entities from a natural world that is otherwise fluid, processual, and undifferentiated. Taxonomists stabilize the attributes of organisms and then use these attributes to distinguish one taxon from another. This activity reduces the noise of nature and distills it into individuated entities. This allows us to socialize the natural world by incorporating it into the linguistic and scientific discourses. Once the work of atomizing nature into discrete species taxa is complete, scientists must then follow up this task with the process of reconstructing those very entities back into an organic whole—into a classification.

Taxonomists make decisions about what goes into a classification system, where it is to be positioned, and what is to be left out, beyond the methodological boundaries of the system itself. Do we lump giraffes into one species, or do we break them up into four distinct species based on DNA testing (Fennessy et al. 2016)? The former conceals the possible existence of three others, while the latter makes it seem as though nature is somehow more wide-ranging in terms of its diversity. Reduction is an important concept in the process of building classifications, and the politics of this selection process is not to be undervalued. In *Counting Species: Biodiversity in Global and Environmental Politics* (2015b, 47), Rafi Youatt looks at the political history of biodiversity from the 1980s through the 2010s, focusing on how the concept of biodiversity has been part and parcel of, and perhaps primarily, a form of global political practice. "I take the position," Youatt notes in relation to bio-census activities, "that the global biodiversity census is as much about power and political life and the boundaries between nature and society as it is about scientific information gathering for conservation ends. . . . Here the focus is on considering the field of social power in which scientific efforts take place, and asking questions about the discourses, resources, and networks that make a biodiversity census possible" (2015b, 47–48). Yet, when people encounter biodiversity classification systems, they often experience them, in practice, as if they are complete.

Our typical understanding of biodiversity is often visual—diagrams and tree-relationships, for example, dominate the domain. Imagine being asked the question, How are dingoes related to domesticated dogs? Even if you know the answer, chances are, at some point, a tree diagram will have become one way you came to understand this relationship. And surely a schematic would need to be made to communicate your vision of the relationship with some other individual. We have been trained to use these representations over the course of hundreds of years. The Porphyrian tree is one of the oldest known tree diagrams, interpreted many times over back to the third century AD. The Porphyrian tree expresses the relationship between Aristotelean genus and species, along with branches diagramming his famous dichotomous divisions (Aristoteles 1995, pt. History of Animals, Parts of Animals, Categories; Lima 2014, 28).

From my early days learning about natural history in elementary school, the "tree of life" image was a mainstay in understanding how species are interconnected and related. Dinosaurs were (and still, in part, continue to be) my jam. I'd labor for hours over different taxonomic charts. Relationship trees are aesthetically pleasing and easy to memorize. But, even before trees, encyclopedias were compiled to make the natural world portable. Pliny, for example, catalogued the known natural world in his *Naturalis Historia* (an original manuscript does not exist, but see 1472 for an example). Regardless of whether a catalogued entity was "real" in any traditional sense (the sciapod category, for example, was a mythical creature included in his original manuscript), Pliny's text allowed the general public the opportunity to peer into distant worlds—he made the distant intimate; the unknown, known, as an element into social consciousness. Linnaeus, similarly, created a nomenclatural pseudo-classification system that became part and parcel of what it means to broadly classify the natural world at all. To have the name *Canis lupus dingo* locates the dingo in the *Canis* genus, the *C. lupus* species, and *C. l. dingo* subspecies.

Originally, Linnaeus's naming system was intended as a technology to facilitate memory and, eventually, a collocating and retrieval mechanism for cogent and easy access to species identification and knowledge aggregation. The resultant nomenclatural system became part of the public imagination, as did the graphical structures that became part and parcel of the depiction of the natural world. These structural metaphors (Lakoff and Johnson 2003, 152) of the natural world are powerful and integrally shape our aesthetic and positional relationship to the natural world. Trees of life are sharply delineated and serve as a mechanism for wresting definition from a landscape defined by complexity. As Lakoff and Johnson note, "Metaphors have entailments through which they highlight and make coherent certain aspects of our experience" (2003, 156).

Classifications do not define reality, however, even if our particular understanding of the "real world" is defined by those same classifications. In fact, as many scholars have noted, the visual mechanics of classifications do influence and define how we negotiate and interpret our social world. All models, interfaces, visualizations, and diagrams—classifications and trees included—interpret underlying data. The resultant images make given conclusions seem intrinsic to the data and, thus, inherent to the world from which they were extracted. As Johanna Drucker notes, such visuals are performative in some respect; she calls them "acts of interpretation masquerading as presentation" (2014b, 10). The allure of the graphical representations of hierarchies has deep roots in the history of the human communication of knowledge (Lima 2014), and their power in the context of everyday settings should not and cannot be overlooked.

Even more insidious and pervasive are the classifications that go unseen amid the daily practices that structure some of our most important resources of general information and knowledge. Technological advances, especially those associated with internet and web environments, often function with the assistance of dozens of overlapping algorithmic classification systems. Classifications are now flattened, fragmented, and distributed throughout complex databases that shape the way we navigate our world. But because these classifications are not easily accessible and graspable through simplistic graphics, they are often not experienced or identified as proper classifications. Yet, they are, and the result is that, while these classifications still individuate and reduce a complex world, we often overlook their effects. Google is a prime example of a tool that uses algorithmic classification to rank, order, and provide access to the countless documents and webpages available on the web. Safiya Noble's (2018) work is paramount here, in that it lifts the veil on the seeming objectiveness of the platform. Noble shows how Google's search results have propagated racist ideology by "technologically redlining" and exacerbating the already present modes of racism prevalent in our society. Noble notes how Black identities, particularly young girls and women, are frequently sexualized and fetishized in online search environments, perpetuating the culture of harm and violence that these same communities experience in social spheres. Technology is nothing if not essentially human by design—design that is impressed by the makers' biases and inclinations.

Such misconceptions stem, in part, from the inevitable and necessary process of reduction. One danger of scientific classifications is that the discourses using them as evidence often assume that they represent nature—if not in full, at least in part. That they are partial and designed is rarely discussed, at least outside the biodiversity world where this reality is an obvious point. Henri Lefebvre writes,

> Reduction is a scientific procedure designed to deal with the complexity and chaos of brute observations. This kind of simplification is necessary at first, but it must be quickly followed by the gradual restoration of what has thus been temporarily set aside for the sake of analysis. Otherwise a methodological necessity may become a servitude, and the legitimate operation of reduction may be transformed into the abuse of *reductionism*. No method can obviate it, for it is latent in every method. Though indispensable, all reductive procedures are also traps. (2011, 105–106; emphasis original)

A trap, indeed, and an inevitable one. Lefebvre's point, in part, is that reduction has the potential to then be subsumed into our mental reality, and thus representations infiltrate the space of daily practice and the imaginative realm. A key to understanding a classification is to understand the methods, rules, and policies that dictate its ultimate form. The danger of the reductionism described by Lefebvre is of two kinds: on the one hand, the scientist can fall prey to the functionality and ease of their reductionist approach such that they become comfortable "curling-up happily" in the warmness of their niche (2011, 107). This, to me, is less dangerous in practice, given that the biodiversity specialists I encountered in my fieldwork were anything but uncritical. They were each well aware of their methodological approach, its pitfalls, and the dangers of artifacts that simplify complex and nuanced work.

This reflexive, critical attitude no doubt emerges in part from the longstanding debates between different taxonomic "factions." On the other hand, the more dangerous effect of reductionism, from Lefebvre's point of view, stems from the fact that the reflexivity exhibited by the biodiversity scientists is not typically practiced outside of the domain. So, whereas biodiversity scientists are fully aware of the limitations of classifications, nonspecialists do not have the expertise to negotiate these structures effectively and to critically understand their pitfalls and nuances. Taxonomic structures, such as they are graphically delivered, simplify and obfuscate the conflict between taxonomic approaches that is readily apparent to those within the discipline. In this way Lefebvre sees this kind of reductionism as a "tool in the service of the state and of power" (2011, 106). Individuals proceed as if these graphical statements are statements of fact and not as if they are hypothetical structures that merit careful scientific consideration.

UNIVERSALITY AND THE MYTH OF TOTALITY

One cannot speak of the reductionary qualities of classifications without also broaching the concept of universality. When we speak of universality in relation to classification, we often mean "this classification organizes, or can organize, everything." Yet, what actually defines universal structures, in practice, is that they simplify complex sets of ideas and express them in a univocal fashion. In his *Quiddities* (1989), Quine, reflecting on the notion of the "Universal Library," highlights the inherent contradiction and absurdity of this "melancholy fantasy." As illustrated in Borges's "Library of Babel" (1999, 112–117), the infinite space of the library is untenable and maddening, where rationality "is an almost miraculous exception." With no comprehensive library index, despondent librarians are useless and unable to unlock sense from this senseless infinity of letters, words, and text. In a move fitting of Ockham, Quine whittles down the infinite to a similarly confounding sensibility, that all this complexity can emerge from the finiteness of Morse code: "The ultimate absurdity is now staring us in the face: a universal library of two volumes, one containing a single dot and the other a dash. . . . The miracle of the finite but universal library is a mere inflation of the miracle of binary notation: everything worth saying, and everything else as well, can be said with two characters" (1989, 42–43). And therein lies the ever-problematic notion of claiming universality: to be truly universal, one needs to represent all knowledge, and yet, in practice, this is an impractical possibility. The only solution is to fall back to the (seemingly) infinite possibilities of language.

For one, if classifications, by definition, reduce the natural world to a series of perceptible and documented tokens (linguistic or otherwise), then how can we *also* purport that classifications are also universal? Using the term universal and speaking of its discursive use in both information studies and biodiversity could cause some confusion, so some clarification is in order. Within the realm of information studies, universal classification systems are those general systems that are able to expand and adapt to the entirety of extent knowledge—that is, systems that are able to classify documents in any and all disciplines. For example, systems such as the Dewey Decimal Classification and the Library of Congress are created in a way that all disciplinary literature can fit within their established scheme, and as such, they are universally capable of expanding and absorbing concepts within its scheme. These types of classifications are often associated with the modernist turn in the information sciences (Mai 2011).

On the face of it, this is a virtuous goal. As recently discussed by scholars in the subdiscipline of knowledge organization (Dahlberg 2017; Gnoli and Szostak 2014), to be universal is to enhance cross-disciplinary conversations and support system interoperability. Some of these same theorists have posited that a universal method based on nature's fundamental properties can more accurately organize knowledge from multiple domains. "The theory of integrative levels claims that the natural world is organized in a series of levels of increasing complexity: from physical particles and molecules, through biological structures, to the most sophisticated products of human thought" (ISKO Italia 2004) (see also, Gnoli and Ridi 2014; Gnoli and Poli 2004; Szostak 2008). This adherence to a static understanding of external phenomena (consistent enough, in practice, that the organization of *all* knowledge can perpetually be conformed to this schematic) is of little use, however, in domains as specific as biodiversity taxonomy, where the "reality" in question is a shifting ground of concepts and taxonomic arrangements. These reductionist approaches not only presuppose a permanence and consistency to our general knowledge that just does not exist, they also overlook the socially situated, culturally defined unfolding of our knowledge production practices.

As we look to biodiversity systems such as the Catalogue of Life, these consensus structures are implemented to facilitate cross-group data sharing and data aggregation from disparate sources. Yet, as demonstrated, a classification not only *organizes* information, it also *creates* very distinct epistemic boundaries that proliferate throughout society in intended and unintended ways. As such, discussions regarding what universality means must attend to not only what it facilitates, in terms of discovery, but also what it inhibits and epistemically distributes via the epistemic commitments that underlie its universal modes of aggregation.

In the biodiversity sciences, universality is used similarly, but also differently, in certain respects. Of course, unlike general documentary classifications systems, the practical function of most biodiversity taxonomies (say, for example, a taxonomy covering all weevils) is not to represent all knowledge, but rather to represent an organism-relationship model that attends to the particular theoretical and epistemic orientations of one scientist or group as it applies to a finite set of entities or concepts. In theory, a taxonomic approach *can* be used to organize all species. But taxonomists specialize and function on small scales. This is partly why consensus classifications have grown so contentious—it is a different matter to organize *everything*, since no one person can accomplish this to any effective degree. This falls in line with Hjørland and Albrechtsen's (1995) articulation of the domain classifications, an approach that Hjørland has vehemently supported for years (2009). Within the boundaries of these species-specific classifications, what is universal is not, necessarily, that the classification itself can include all species. What is emphasized as universal is the particular methodological approach to articulating species and their relationships. If a scientist uses a cladistic approach, for example, to organize weevils, that same approach can be used to examine any segment of the animal kingdom. But, even more than that, when you listen to the discourse of biodiversity scientists, to be universal actually means to have a classification method that can be of use to all scientists as a mechanism of construction. In some spaces, the assumption behind a *truly* universal system is that it represents a "natural" organizational principle that can be used as a baseline against other methodological approaches (Doolittle 1999b).

As Mai has noted, the impulse of "modern" classification systems to standardize and universalize knowledge organization systems around notions of "exclusiveness" and "exhaustivity" overlooks the application of standards in individual domains (1999, 548–551). In recent years, a more historically informed notion of classification has taken shape, one in which "unificationism" has given way to a "generation of theories, principles and methods that emphasize both the cultural and historical specificity of classification practices and their emancipatory function" (Furner 2013, 32)—a topic we attend later in this text.

INTERNAL EPISTEMIC PRACTICES

Expanding this, what does it mean to "uniformly implement a method" in a taxonomic space? In short, most taxonomies are built along a set of uniform rules. By following a rule throughout the taxonomic space, a taxonomist can validly argue for one particular arrangement over any other. Understanding these rules is precisely what Patrick Wilson was referring to when speaking of bibliographical systems as instruments. And, this instrumentation is important to understand, especially, with respect to the Catalogue, how we can critique its shortcomings. Along with every subsidiary taxonomy the Catalogue collects comes a particular taxonomic structure that represents the opinion of the experts that created it. "The source databases are diverse in their origin, their purpose and therefore their structure. A key challenge for the Catalogue of Life has been the integration of this disparate data, and a standard dataset has been established for that purpose" (Species 2000 2015c). The distinctions between one condition and another represent deep-seated theoretical and philosophical divides about how relationships can and should be built within classification systems, and how species are lumped into nested and hierarchical taxa.

An in-depth explanation of biological taxonomy is outside the scope of this book, but I can broadly describe some examples to show how different schools of thought go about creating taxa and connecting them in networked relationships. Marc Ereshefsky identifies evolutionary taxonomy, cladism, and pheneticism as three possible ways to construct a biological taxonomy (2007, 50–51). Evolutionary taxonomists believe that the emergence of new species taxa can occur through two distinct processes: cladogenesis and anagenesis. Cladogenesis is the splitting (branching) of a "single genealogical lineage" (2007, 52), through, for example, the process of occupying new adaptive zones (a population of a species becomes geographically isolated from the rest of the population and adapts with new genetic characteristics) (SæTher 1979, 308–309; Ereshefsky 2007). Anagenesis, on the other hand, is the gradual change (divergence) (Ereshefsky 2007, 52) over time of a species until it becomes distinguishable as a new species. The mechanisms and qualities used by taxonomists to assess "significant" (Ereshefsky 2007, 52) enough changes to warrant a new species for anagenic change is a subjective process (Vaux, Trewick, and Morgan-Richards 2016) and a source of much debate in the taxonomic arena. Cladogenesis produces monophyletic taxa: taxa that include an ancestral species and all of its descendants (Grant 2003). Anagenesis, however, produces paraphyletic taxon groups: taxa that contain some but not all of the descendants of a particular taxon (see figure 5.1).

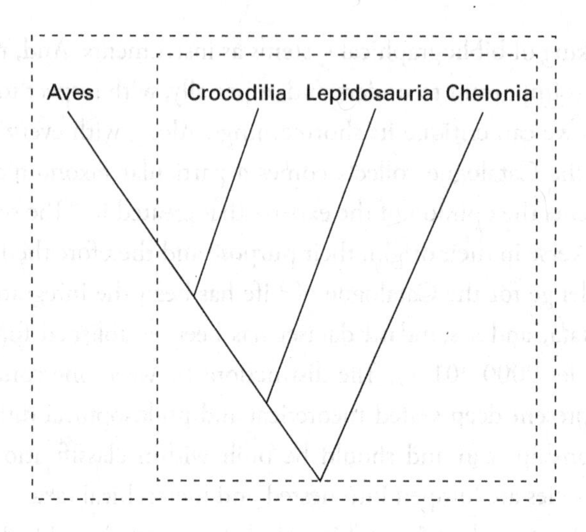

Figure 5.1
Example of monophyletic (outer box) and paraphyletic groups (inner box). Evolutionary taxonomists would define the class Reptilia as containing only lizards, snakes, and crocodiles, excluding Aves as part of this schematic; thus, it is paraphyletic. Cladists, on the other hand, would be unwilling to exclude Aves, because groups should include all the descents of a taxon, and thus a Reptilia group that includes Aves is monophyletic. Figure adapted from Marc Ereshefsky's *Poverty of the Linnaean Hierarchy: A Philosophical Study of Biological Taxonomy* (2007, 54) and "The Reptile Database" (Uetz 2016).

One might look at figure 5.1 and ask why this seemingly minute distinction matters. After all, the tree is essentially the same, regardless of whether you understand the class Reptilia to contain birds or not. What differs is the way each school *interprets* this diagram to represent some argument about how animal groups form taxa. However, while the differences seem subtle in figure 5.1, stating that these taxa occupy different class *positions* means they occupy vastly different spaces in the nested taxonomic hierarchy of the Catalogue database. In figure 5.2, an evolutionary point of view (according to Ereshefsky's model) situates birds (Aves) as a different class. Theoretical and interpretive distinctions amplify themselves in online taxonomic structures.

Cladistics, in contrast to evolutionary taxonomy, does not accept paraphyly as part of their construction of taxa.[2] Cladists would balk at separating Aves from the Reptilia group since cladists base the construction of their taxa on the concept of shared traits and common ancestry (Ereshefsky

▼ kingdom: Animalia
- ▶ phylum: Acanthocephala Rudolphi, 1802
- ▶ phylum: Annelida
- ▶ phylum: Arthropoda
- ▶ phylum: Brachiopoda
- ▶ phylum: Bryozoa
- ▶ phylum: Cephalorhyncha
- ▶ phylum: Chaetognatha
- ▼ phylum: Chordata
  - ▶ class: Actinopterygii
  - ▶ class: Amphibia
  - ▶ class: Appendicularia
  - ▶ class: Ascidiacea
  - ▶ class: Aves
  - ▶ class: Cephalaspidomophi
  - ▶ class: Elasmobranchii
  - ▶ class: Holocephali
  - ▶ class: Leptocardii
  - ▶ class: Mammalia Linnaeus, 1758
  - ▶ class: Myxini
  - ▶ class: Reptilia
  - ▶ class: Sarcopterygii
  - ▶ class: Thaliacea
- ▶ phylum: Cnidaria
- ▶ phylum: Ctenophora
- ▶ phylum: Dicyemida
- ▶ phylum: Echinodermata

Figure 5.2

Catalogue of Life 2016 Annual Checklist taxonomic tree depicting the separate placement of the class Aves from class Reptilia in the tree structure (Species 2000 2020). CC-BY 4.0, Catalog of Life, used by permission.

2007, 55). Cladists reject the decision to separate birds from the Reptilia group because there is no hard and fast way to delineate when one species split off into another.

Phenetics, our final approach under discussion, produces a distinctly different representational hierarchical arrangement. Numerical taxonomy, or phenetics, is a probabilistic method that groups organisms together based on the general premise that those with the most phenotypic overlap are necessarily more closely related. As articulated by Robert Sokal,

> Numerical taxonomists contend that evolutionary importance is undefinable and generally unknown and that no consistent scheme for weighting characters before undertaking a classification has yet been proposed. To weight characters on the basis of their ability to distinguish groups in a classification . . . is a logical fallacy. Since the purpose of employing the characters is to establish a classification, one cannot first assume what these classes are and then use them to measure the diagnostic weight of a character. (1966, 109)

Numerical taxonomy was touted as the apex of empirical science when it burst onto the scene, particularly because such analysis, in theory, would always produce the same result from laboratory to laboratory over time (Sneath and Sokal 1973, 11). Codes were created for each particular object trait: "'hairiness of a leaf' might be coded as follows: hairless: 0, regularly haired: 2, densely haired: 3" (Sokal 1966, 114). Of course, as is the case with any method deemed empirical, one must critically assess the choice of variables or characteristics that undergo analysis for clustering. A notable weakness of numerical taxonomy is the fact that a relationship between organisms is defined solely by "similarity . . . operating on the assumption that the total phenotype accurately effects genotype. [Numerical systematists] believe that an unweighted measure of overall similarity provides an accurate determination of relationship. In so doing, pheneticists ignore the possibility of evolutionary convergence" (a circumstance where unrelated organisms evolve similar traits because of environmental influence) (Pietsch 2015).

Every classification is but one way to represent a phylogeny that is altogether too complex to be represented in any one graphical structure (Hull 2001, 227)—there are far too many variables (known and unknown) involved in the process of evolution. As D. L. Hull states, "Any one phylogeny can be classified legitimately in many different ways. . . . Only the most generic and impressionistic inferences about phylogeny can be drawn from an evolutionary classification" (2001, 227). Regardless of which method of taxonomic arrangement prevails as dominant in our contemporary climate, elements of these schools demarcate sharp differences in the way classifications are constructed and interpreted as modes of argumentation. What matters most is that each classification produced by any of these methods, or any other, must adhere to ontologically and epistemically informed rules; these metaphysical and epistemic commitments then permeate the entire taxonomic structure.

The Catalogue, however, unlike these description-based approaches, is not concerned with any one approach to producing taxonomic knowledge. Each section of the Catalogue's management classification might adhere to different commitments depending on the source of the taxa in question. For example, the editorial group for the management classification of fungi might choose one "master" taxonomic approach, whereas another authoritative approach may be implemented for birds. In other words, the Catalogue is not uniformly internally consistent, which is a major issue for many in the taxonomic field.

RETRIEVAL OR DESCRIPTION?

A pertinent question here is, What are biodiversity taxonomies attempting to accomplish, particularly in light of the uniform application of method to any given instrument? That is, what is their function from the standpoint of *use*? Jonathan Furner proposes one way that we can potentially assess classification systems:

> I think it is possible to distinguish two conceptions of the goal of the practice of KO [knowledge organization], and this distinction corresponds roughly to the one Raya Fidel draws between two conceptions of the goal of indexing. On the one hand, we have the document-centered view that indexers should aim to assign index terms to documents (or documents to index terms) in whichever way it is that produces the most accurate representation of that content. On the other hand, there is the user-centered view that indexers should aim to associate documents with those terms that are most likely to be used by searchers looking for those documents. Similarly, I think that, on the one hand, we have a description-oriented conception of the goal of KO, being to build systems that do well at helping people produce accurate descriptions and representations of documents. And on the other hand, we have a retrieval-oriented conception of the goal of KO, being to build systems that do well at helping people find the documents they think they want to find. (2009b, 9)

This distinction between description- and retrieval-oriented approaches to classification seems to me a possible way to categorize the two basic kinds of taxonomies that currently flourish within the biodiversity world. As Tony Rees, manager of the Divisional Data Centre, CSIRO Marine and Atmospheric Research, in Tasmania, articulated this on the Taxacom biodiversity listserv: "I look upon biological classifications as serving two purposes—first, to illustrate our current best guess/es as to the relationships between organisms, and second, to provide a recognisable navigation structure so that persons entering the classification can (hopefully) find their way to their particular organisms of interest" (2009). This quote brings to light the two distinct aspects of the biodiversity instrument. For Furner, these two approaches are rooted in the larger question about how one is to *evaluate* classification systems as systems that "represent relationships of identity between classes of documents," and "help people find the right labels for classes of documents that about those identities, and help people find those documents" (2009b, 4).

Although there exists no standard by which the true "goodness" of a classification system can be quantitatively assessed, a number of variables can be identified to critique the effectiveness of classification schemes. Furner indicates that the basic role of a classification system is to adequately represent the identity of some external reality: "The main claim that I would like to make about the importance of identity for KO is not that an understanding of identity is helpful in analyzing the structure of aboutness and relevance. It is that there is a sense in which identity is actually the goal of KO" (2009b, 12). The rubric for assessing whether classification systems "work well" can be distilled to a series of factors, from within either a "description-oriented" or "retrieval-oriented" notion of classification (2009b, 9–10).

According to Furner, description-oriented classifications ask two basic questions of a designed system: (1) How correct, just, and fair, is any given ontology in relation to the natural, real world? and (2) How internally coherent is the infrastructure itself in exemplifying a unified ontological system with an internal logic? (2009b, 9). In the biodiversity realm, this means creating classifications that provide a consistent model that includes a fair and accurate representation of biological organisms, as well as one that provides a classificatory system that depicts things the "the way things really are, or the way somebody thinks things are" (2019, 9). Traditional, internally consistent taxonomies can be defined as functioning in this way; they are systems that invoke a unified methodology throughout, and thus, provide a consistent model of the natural world. Because descriptive systems are consistent, a species location in a system tells you about how it operates in relation to all other entities within that system. In terms of the Catalogue, we might call most of the contributed taxonomies descriptive. This was certainly the case for our previously discussed WeevilBase, for example, as it was brought into various global infrastructure. A standing question that remains in this case, however, is how we define "reality" as a goal of a descriptive system when nature is characterized by change. Entities evolve, and so our classifications will always be catching up to this moving target.

Retrieval-oriented classifications, on the other hand, are evaluated by way of a classification's ability to facilitate the locating of documents or required information by a user (2009b, 10). These systems, in theory, aim to model classifications in ways that are expected, or understood, by users. Terms Furner associates with this level of assessment include effectiveness, efficiency, and usefulness to the user. In the biodiversity world, this would mean creating a system that has the ability to adjust to meet the expectation of many kinds of users and points of view: "different people see reality in different ways," says Furner (2009b, 9). Such a view prioritizes variables other than description as the core role of a system. The Catalogue, for example, is attempting to use consensus to find an epistemic middle ground that meets the needs of many user expectations and applications. It is thus more pragmatic. Evaluating from a retrieval standpoint, some might say that the Catalogue is more effective because it is more comprehensive and global, even if the internal taxonomies might contradict one another—that it is more efficient because it is a one-stop shop, and more useful because it can also serve as a backbone taxonomy.

My intention in making this somewhat fluid distinction between description- and retrieval-oriented classifications is to bring attention to two broad approaches to classification in biodiversity sciences: one is the product of individualized scientific work and hermeneutic development, and the other is a space that attempts to prioritize consensus to unify information access and facilitate communication. This boundary tends to mark the divide between those that support and those that do not support a generalized taxonomic model such as the Catalogue of Life. To be sure, Furner also notes that the division between description- and retrieval-oriented approaches is artificial, for these approaches often commingle in practice. Historically, Linnaean nomenclature—terms that with their genus-species designation gesture toward a general sense of rank—were created for memory retrieval, after all. The trick, as with all retrieval systems, is to balance correctness (however that is defined within a certain context) with facility of use (for expert and nonexpert audiences alike).

The biodiversity taxonomic instrument is a complex machine. As we have seen, systems necessarily reduce nature to distinct entities and, in the process of creating classifications, must consistently reassemble these entities in ways that adhere to a certain set of epistemic commitments. From a process-oriented perspective, this is problematic, as it erases the system-oriented context from which these entities are derived. Nonetheless, classifications are practically necessary if we are to communicate effectively about nature and manage our relationship to it. The problem is that there are many ways to communicate this information. Taxonomic commitments vary widely in practice, and they are often fundamentally incommensurable. One problem with rigid, description-oriented taxonomic systems is that they are univocal and do not represent, or pretend to attend to, multiple points of view. This is, in part, because these descriptive taxonomies serve hypothesizing functions—they are arguments. The benefit of them, of course, is that they maintain the intellectual integrity of those who built them.

But even while these descriptive taxonomies may be incommensurable, structures like the Catalogue of Life are nonetheless trying to commingle these taxonomic opinions in spaces that represent consensus. The assumption is that, by aggregating multiple sources of data, retrieval-oriented concerns can be met. However, given the traditional, epistemic functions of classifications, these systems run counter to taxonomic expectations. In chapter 6, we switch our focus to the space of the Catalogue in order to better understand its intended benefits, even while its ultimate function is up for debate.

6 INSTRUMENTAL POWER

> If we do not know the extent of the power that a particular instrument gives us, we shall not know what to make of our apparent failures, if there should be any, and shall in fact be unable to distinguish failure from a sort of success.
>
> —PATRICK WILSON
>
> *Two Kinds of Power: An Essay on Bibliographic Control* (1968, 35–38)

Given the described differences between traditional, description-oriented classifications, and composite-based, retrieval-oriented management classifications, one might ask the question, What are the *emergent* properties of these composite systems as they pertain to the biodiversity ecology? As the epigraph by Patrick Wilson indicates, the kinds of power we have over systems must be unpacked if we are to understand the extent of potential control we have over the documentary universe in question. To extrapolate on this potentiality, I outline below the kinds of instrumental power that consensus systems embody in the domain of biodiversity studies. To better understand this power, I expand on two related questions: (1) How do these systems allow us to better understand the extent and limitations of our current biodiversity knowledge? and (2) How does such functionality influence the way biodiversity knowledge is interpreted on a distributed basis, in terms of structuring data entities as well as allowing for the generation of new forms of inquiry?

To answer these questions, I introduce the concept of taxonomic *extensibility*, which I define as the extent to which consensus structures shape biodiversity knowledge in specific, targeted ways. These kinds of emergent potentials are often, but not always, invoked in domains outside taxonomy proper. Such extension functions both within the Catalogue's own taxonomic space and within external systems. *Internal extension* relates to the Catalogue's ability to identify and compensate for certain gaps in its taxonomic record; that is, by bringing together global data, the Catalogue is able to identify the general contours of the biodiversity data landscape. Taxonomic gaps occur for many reasons, but typically they define areas that contain uncharismatic species (those poor mussels in the Bivalvia class, for example) and underfunded geographies in biodiversity research (such as areas in rural Africa). To bridge these omissions, the Catalogue has introduced the concept of a proto-GSD, which gathers records from many unrelated sources to construct a tentative and intermediary database to temporarily overcome these taxonomic shortfalls.

Management hierarchies are also used to shape external biodiversity databases and to produce new forms of speculative knowledge. These external databases, such as Global Biodiversity Information Facility (GBIF) and the Encyclopedia of Life (EoL), are heavily influenced by the Catalogue's internal and editorial interpretations. In this chapter I outline how one structure, GBIF, integrates the Catalogue into its database structure. The materiality and epistemic influence of these backbones are then explored. Finally, we briefly illustrate some more generative, extensive functions of these hierarchies: their use as knowledge bases intended to predict evolutionary futures based on networked biodiversity data from the past. Such speculative spaces extend the consensus structure's epistemological virtues to its furthest extent, embracing the ever-contingent nature of taxonomic work as a foundation for new modes of hypothesizing work.

This extensibility, I argue, has implications for how we can conceptualize the potential of documentary classification systems in general and how we can reposit how they are positioned as epistemic instruments in information studies.

EXTENSIBILITY

What affordances do consensus structures provide with regard to the production of biodiversity knowledge at the global level? What might we be able to say about the extensibility of the Catalogue versus the extensibility of any other taxonomy? As mentioned in chapter 5, the Catalogue's commitment to access produces an operational and facilitative structure—it helps people get things done. I call what consensus taxonomies can get "done" their *potentialities*, or their *extensive capacities*, and the power they provide to the user (any one user or group of users) an *instrumental power*. In short, consensus taxonomies expand on existing taxonomic knowledge to produce emergent knowledge that exceeds the sum of the taxonomies included in the system. The extensive taxonomic knowledge they produce is understood to be imperfect and is often speculative and temporary.

Looking back to Patrick Wilson's *Two Kinds of Power* (1968), we see that he presents two powers that exist within the bibliographical universe: descriptive power, defined as the ability to describe documents in such a way that we can call up a set of undifferentiated, related documents; and exploitative power, which he describes as the ability to use texts in a manner most relevant to a situational circumstance. A quality of both powers is that they measure what can be performed with documents *within* a system. Further, these powers are predicated on the articulation of a bibliographical instrument (the system) that has particular specifications designed to systematize access to, and maximize, said powers (Wilson 1968, chap. 4). These specifications include the articulation of an instrument's domain and its bibliographical (or documentary) extent, descriptive capacities, logical transparency, and institutional context, among others. The more an instrument meets these ideal instrumental specifications, the more power it affords a user with regard to accessing and using *documents*. The extensibility of a consensus structure, or its instrumental power, similarly depends on these specifications, but they expand on Wilson's original schematic in that this extension focuses on the exploitation of (and access to) the taxonomy system itself, in addition to the documents it contains. Wilson's notion of bibliographical control is concerned with a specific collection of things (the bibliographical universe), and how we can best situate access to these things. The extensive capacities of a system, on the other hand, tell us how the *system* can be exploited in distinct ways—as a kind of über document—to influence the conceptualization of other biodiversity systems and produce new forms of taxonomic and documentary knowledge.

Seeing consensus spaces more as standardized, repurposable, temporary schema, rather than as composed argument, is an important move for knowledge organizing practice, particularly in the biodiversity sciences. This frame shift redefines the primary assumptions associated with scientific taxonomies—most notably, that they be internally consistent and epistemically coherent. This does not, of course, overshadow the downsides of such a shift (and there are, indeed, many!), which we discuss in chapter 7 and 8. Not only do consensus taxonomies organize classes of things, they also serve as overt interpretive frames that becomes embedded in, and deeply influence, the role and future intentions of global biodiversity work. The flexibility they afford is built into their construction, as part of their design. These instruments provide a structure for other systems that, in practice, actively rearticulates and re-presents the Catalogue's core classification through the lens of its own classificatory expectation and needs—a veritable onion layering of epistemological interpretations.

This kind of extensive activity seems to me appreciably different from the expectations placed on classification systems typically seen in information studies spaces. The Dewey Decimal Classification (DDC) and the Library of Congress (LoC), for example, were built as systems to guide the organization of documents within multiple physical (and digital) locations. The supposition is that these structures are to be centrally controlled to facilitate class consistency, as well as to limit the extent and speed in which the system changes over time. And despite the DDCs "vigorous" policy for change, the rate of change represented in these spaces comes nowhere near that of taxonomies in biodiversity work (Miksa 1998, 25)—and with the rise of genetic taxonomic approaches, we will see the rate of this change speed up dramatically in the future. Further, as systems like the LoC and DDC change, these changes are theoretically expected to trickle down, so to speak, to local repositories, so that they can remain current with the schedule (we will ignore, for the time being, that public libraries, especially, do not often change book positions on their shelves when such changes occur) (Miksa 1998, 27). On the other hand, while the Catalogue is designed for integration in external systems, its creators understand that those systems may, or may not, ever upgrade their core architecture as the Catalogue evolves, even as the Catalogue continues to make such upgrades easier through the development of persistent identifiers and application programming interfaces. A taxonomic backbone is often seen as a starting point for some external system.

One might argue the position that any classification instrument can be extended. Somebody can, for example, take the DDC classification schedule, download or replicate it, and implement it for their own local purposes, manipulating it as they see fit (aside from whether or not permission was granted or needed for such reuse). This is true, of course; anybody can perform this activity as a kind of imposed power onto the classification instrument. I can take the DDC schedule and do what I wish with it—there is nothing stopping me from doing this. But differentiating this kind of imposed use as categorically different from the use of any document for any other particular purpose would be a fruitless enterprise. An imposed repurposing is not unique to the Catalogue or any other instrument, so it does very little to help us articulate the power of extensibility. The difference lies in the fact that consensus systems like the Catalogue take this repurposing as a core value of their access-oriented commitments, and design the system to facilitate these possibilities. The Catalogue *actively* builds extensibility into its functionality—it is *designed* to be extended into other digital and organizational domains, and so the practices, standards, and policies that arise from within the organization are in support of this facilitative activity. One value that supports this kind of extensive capacity, for example, is that the Catalogue is not trying to control the internal consistency of its taxonomy—conflicts between classes is understood as a possibility, and maybe even an inevitability. Systems like the DDC and LoC depend on classificatory control to maintain consistent descriptive systems over time. A key value of the DDC, for example, is that its classificatory approach must be as consistent as possible over time, lest schedules conflict with one another. For a local repository to take the DDC classification system and amend it in any number of ways somewhat defeats the original bibliographic and documentary function of the DDC. If a local repository strays too far from the classification standard, updates or revisions to the main classification schedule become irrelevant to the local collection, and thus inhibit the ability to control and exploit the system in the future.

The concept of *extensive power* helps us understand the potential functionalities consensus taxonomic instruments have within digital spaces, as well as their powerful mediative functions. Digging more deeply into the extensive capacities of backbone taxonomies, let us now examine the levels at which new forms of knowledge are produced.

Internal and External Extensibility

The extensibility of consensus taxonomies functions on two distinct levels: internal extension, which focuses on knowledge contained and produced within the consensus taxonomy itself, and external extension, which speaks to the effects on knowledge these structures have on or within external systems (figure 6.1).

First, with regard to internal extension, it is sometimes the case that authoritative global databases are not available for a particular taxon, or

The taxonomic instrument

Instrumental power: Over documentary instruments; extensive capacities of organizational structure; combinatory possibilities

Internal extensive capacities	External extensive capacities
- Identifying taxonomic gaps (either geographic or taxonomic) - Assembling taxon groups through data aggregation and inference	- Serving as backbone for external data- and knowledge-bases. - Predictive activities supported by consensus structures

Figure 6.1
The extensive capacities of taxonomic instruments.

data is unavailable for certain global geographies. For example, imagine that a singular authoritative taxonomy did not exist for the genus *Ursus*. Yet, data on *Ursus* may be available in other locations, distributed sporadically in many different contributed taxonomies. In these cases, data is aggregated from these disparate systems and brought together to achieve adequate, or near-adequate, coverage for that particular taxon or area. To best create the most complete taxonomy for *Ursus*, some information might be collected from databases in Mexico, Russia, and Spain to create a semi-complete aggregation of knowledge for that taxon. The resulting internal structure is imperfect at best, to be sure, but yet it still provides a tentative taxonomic view (or, at least, backbone) that allows for the organization and integration of data that exists about *Ursus*. Internal extension, then, is the extension of the knowledge contained *within* the consensus taxonomy itself toward the end of more in-depth and comprehensive taxonomic coverage.

External extension functions in two ways. First, consensus systems provide a data structure (the management taxonomy or taxonomic backbone) and classificatory foundation for organizations and scientists seeking to organize data in online environments. The Global Biodiversity Information Facility, for example, is one global organization that builds on the Catalogue's taxonomy to create a larger data structure outfitted to its purposes. In situations such as these, consensus structures influence the shape and constitution of knowledge in external environments, radiating access-oriented commitments (and the myriad assumptions the edited system represents) throughout the domain of biodiversity work. Second, and perhaps more radically, consensus taxonomies are seen as vital instruments in the emerging discipline of evolutionary informatics, which takes collected information and attempts to answer evolutionary questions and predict ecological and evolutionary trends. External extension is thus the radiant and widespread influence of consensus structures on shaping and modeling the knowledge landscape within the discipline of biodiversity, broadly conceived.

If we are to truly understand the long-term effects of consensus taxonomies on the practice of science, we need to understand how they are implemented to facilitate scientific and knowledge-generating functionalities. I focus our discussion on a select few examples of these uses. Of course, these structures have many other potential uses—both known and yet to be discovered—but I highlight the most salient uses uncovered in fieldwork conversations. I begin with the internal extensive potentials of consensus structures and, specifically, how taxonomic backbones are used to help survey the geography of biodiversity knowledge and to subsequently fill identified gaps in the taxonomic record. I then broaden my purview, to how these structures influence the articulation of biodiversity knowledge far beyond its systemic borders.

INTERNAL EXTENSION

Understanding the breadth and depth of current biodiversity knowledge is central to the recent articulation of consensus structures. Aggregated structures allow biodiversity scientists a bird's-eye view of the biodiversity and taxonomic knowledge known to science. Such a view affords the ability to find gaps, errors, and overlaps in global data. As previously noted, Mesibov (2010) illustrated the extent to which database errors proliferate throughout the data ecology. Such identifications help scientists fix these errors across a broad swath of infrastructure—a process that would be far more onerous (if not impossible) if attended to at the local level many times over at each database site. Much more will be said of error proliferation—which certainly presents a significant data management problem in its own right—but it should at least be noted that identifying any problem is the first step toward managing its impacts on integrated systems.

The identification of taxonomic gaps is another significant potential of consensus structures. As contributed taxonomies are brought together, what emerges is a partial representation of the natural world: that part of the natural world that has been prioritized by systematic scientific investigation. The gaps that inevitably present themselves arise from two different, but related, problems: understudied taxa, and underrepresented geographies. Regarding the first problem, more-charismatic species—those species that get more attention from scientists and policy makers (Bowker 2008, 146)—have a tendency to get described and classified more exhaustively. Entwistle and Dunstone (2000) show how 70 percent of conservation funding applications to Fauna & Flora International, a global conservation organization, were for large mammals and birds—taxa that have historically received a great deal of attention. These species then come to symbolize organizational efforts (such as those of the World Wildlife Fund and Save the Tiger Fund), which, in turn, funnel more resources to these species over time (Entwistle and Dunstone 2000). Certain other species groups, on the other hand, such as worms (Annelida) and mollusks (Mollusca), are only sporadically described in biodiversity literature (Kunze, Didžiulis, and Roskov 2013), meaning that they are underrepresented by scientific examination, broadly conceived. In rarer cases, intellectual property disagreements regarding user agreements, data reuse practices, and intellectual control cause taxonomies covering large taxonomic groups to be excluded from the Catalogue. Once a taxonomy is incorporated into the Catalogue, the creator of that database has no control over how the data will be used, which some find (understandably) problematic. Because databases are the result of extensive and long-term individual labor, and because one person (or group) often controls the database, an entire segment of the tree of life is excluded from the Catalogue's global taxonomy if minute disagreements arise from the Catalogue's open-source approach to data management. The result of this kind of omission has the same operational outcome: these species do not get their fair share of representation.

Underrepresented geographies pose still another nuanced problem for systems like the Catalogue. Such underrepresented areas often coincide with areas with unstable or very little local scientific infrastructure; such is the case with certain African geographies that have increasingly become the focus of Catalogue data collection efforts. And although select areas of the southern hemisphere, such as Brazil and New Zealand, have well-funded data collection and animal monitoring programs, the data is still sparse in comparison with more robust species and ecological documentation programs in the northern latitudes (Chambers et al. 2017). Additionally, international conservation efforts tend to favor areas of the globe with large numbers of endemic species that are acutely threatened, also known as "diversity hotspots," which tends to overemphasize these areas in research and data (Conservation International 2017). These hotspots, as noted by

Conservation International, represent over 40 percent of bird, mammal, and reptilian species around the globe, and represent significant portions of Central America, South America, the Mediterranean, and Southeast Asia (Critical Ecosystem Partnership Fund 2017), which means upward of 60 percent of species live outside these zones.

On top of these disparities, the Catalogue must also contend with the species and geographic biases of the Global Species Databases, in that they tend to overrepresent Western geographies. For example, GSDs, which typically focus on one taxon worldwide and contain a taxonomic checklist of all species within it (Species 2000 2015c), are focused primarily in European and North American countries and, as such, tend to amplify species in areas they have easy access to. One reason for this Western focus is that taxonomies and species checklists arising from these areas are bolstered by an integrated and historically strong network of biodiversity institutions and natural history museums—institutions that hold the type specimens and literature central to species identification and descriptive processes (Roskov 2016c).

Historically, countries such as China, Russia, and Brazil have taken a different approach to species data collection, focused on national checklists and taxonomies rather than information that strives for a global reach. The result of these more insular collection efforts is a Regional Species Database (RSD), which poses a special series of problems for consensus systems like the Catalogue.

To reiterate, the lion's share of data ingested into the Catalogue of Life comes from GSDs—these databases provide in-depth and far-reaching global coverage of specific taxa (even if a particular taxa's range is not very large). A genus's range might be relatively small, for example, so a global database in this case does not mean that the species are found throughout the world. These databases are professionally vetted and relatively complete, given their descriptive and analytic focus. But not all species are covered to such depth in GSD environments, nor do GSDs have access to each and every part of the globe. RSDs, then, can serve as potential data sets to fill the gaps left after the collocation of GSD data. The Catalogue produces an intermediate taxonomic space composed of RSD data, which they call proto-GSDs. These proto-GSDs account for these taxonomic gaps, but two major problems present themselves when integrating RSD data into spaces dominated by GSDs: there is often overlap with GSD sources, meaning reconciliations must be debated, and RSDs cannot ensure full and equal global coverage for all species represented in the database, meaning that gaps, to a certain extent, often still persist.

By conglomerating RSDs, proto-GSDs attempt to manufacture taxon lists to the best extent possible given available information. To accomplish this, editors will often compose taxa through both automatic and manual efforts. Such activities, however, present some major complications, such as conflicting taxonomic orientations, duplicated taxon names, and the inevitable conflicts between species concepts of the same name. As covered in chapter 4, two exactly similar names can reference an entirely different set of literatures and type specimens, meaning their circumscriptions can be in conflict with each other (Matthias 2013). This concept differentiation problem is evidenced in the following excerpt from the Catalogue of Life blog:

> Take for example the Family Gentianaceae in the Plant Kingdom. This family of plants has an estimated 1650+ species worldwide. We have two current suppliers of Gentianaceae to the Catalogue of Life—[the] ITIS Regional database and Catalogue of Life China. Together they supply the Catalogue of Life with 552 species in addition to 82 infraspecific taxa. The species *Gentianella acuta* (Michx.) Hultén appears in both checklists where it is a synonym in ITIS Regional and an accepted name in the Catalogue of Life China. This is because some of the species of Gentianaceae are cosmopolitan (i.e. present in North America and China) and the taxonomic concept (i.e. accepted name or synonym) is different. To combine the datasets the Catalogue of Life editors had to resolve these issues before publishing it in the Catalogue of Life. Gentianaceae is now part of a "proto-GSD." (Matthias 2013).

Such taxon negotiations only serve to complicate an already complex consensus space. First, how does one decide if *Gentianella acuta* (Michx.) Hultén is a synonym or an accepted name? This partly depends on the currency of the source database and its reputation. But this process cannot be automated, since each conflict is idiosyncratic and requires expert taxonomic opinion. Second, moving taxa from one area of the taxonomic tree to another is not a simple editorial task. Recall the process of establishing nomenclature and how, to a certain extent, these names reflect taxonomic *positions* (the first part of a name indicates a generic position, while the second part a specific). If a species is moved into another genus, the first part of the binomial name needs to be changed to reflect this modification. If the Catalogue were to attempt to do so, the new name would qualify as a "new combination" (GNA 2020a) and, to be accepted by the taxonomic community as a valid nomenclatural act, it would need to be published in accordance with the particular code governing that group. Aside from causing undue confusion in the taxonomic community (changing a name for merely consensus purposes rather than for descriptive purposes would be antithetical to normal nomenclatural practice), the process would be incredibly time-consuming and expensive. Not practical and outside the operational bounds of the Catalogue even if the resources were available.

The results exhibited by proto-GSDs are promising, even if the aggregation is fairly primitive at this point and not yet scrutinized by expert communities (Roskov 2016c). At minimum, these proto-GSDs tentatively allow the collocation of taxon data that would otherwise be completely excluded from global systems. These mechanisms effectively increase the Catalogue's instantiative power (as in, it has the ability to instantiate species concepts that would have otherwise been silenced in the documentary realm). In some cases, taxa once represented at only a 10 percent level in the Catalogue have been increased to about 50 percent coverage with proto-GSD merging. These statistics are not insignificant, especially for a taxon that is understudied and perhaps, even ecologically threatened. Such inclusion could promote more conservation efforts or more research on a particular taxon. Of course, editors are limited in terms of their own taxonomic expertise, and given that RSDs straddle many different taxa, proto-GSDs are far less authoritative than the GSD segments of the hierarchy. This leads to a great deal of understandable skepticism for biodiversity taxonomists who consider this kind of analysis both misleading and partially researched. And within the bounds of the Catalogue, the limitations of these proto-GSD spaces may not be fully visible or understandable.

Proto-GSDs are difficult to build. Algorithmic software designed to flag taxonomic conflicts and recommend master classifications between overlapping structures are promising and have the potential of releasing this burden from individual taxonomic editors. Taxonomy alignment software such as the Euler Project are also gaining in credibility and use (Chen et al. 2017). Larger backbone systems, such as those built by GBIF, enable algorithmic reconciliation of both names and taxonomic ranks (Döring 2015). Yet, despite early attempts at building this automated infrastructure, such software mechanisms never materialized as part of the Catalogue's workflow, partly because of the initial commitment to a more manually curated taxonomic structure. However, as of 2021, no specialized software or interface has been created with which the Catalogue can perform these aggregating tasks. All the editing work is being done directly within the database, using the relational database itself (Roskov 2016c). Often, doctoral students or postdoctoral academics are performing this proto-GSD work, but this approach is costly, and such temporary work is difficult to sustain in the long run, given its dependence on soft money allocations (Schalk 2016b). More often than not, manually building proto-GSDs is undertaken by the Catalogue's executive editor.

Despite these proto-GSD efforts, there is no aggregative practice generally accepted by all taxonomic specialists. Many "professional taxonomists do not appreciate any kind of technical exercises," such as automatic taxonomic building, Yuri Roskov indicated (2017). "For example, a plant list . . . is being built using software by [the] Missouri Botanical Gardens . . . where they are trying to merge regional floras. I spoke to professional taxonomists . . . and they are very much skeptical about this work. It is a very political process" (2017). For the Catalogue, such interventions are better than nothing, however, since its structure is aimed at temporary snapshots of extant knowledge that will inevitably integrate input and get refined over time. It boils down to the Catalogue's commitment to access and the belief that temporary taxa are better than a structure with gaps.

If the goal is to get an up-to-date snapshot of the world's biodiversity in one coherent structure, the Catalogue will do all that it can to achieve this coverage, even if the mechanisms are temporary and imperfect. A major difference between *taxonomic opinion* and *management hierarchies* is this ability to accept flexibility and contingency as part of an epistemic foundation. The Catalogue's taxonomy is a practical tool that understands and embraces its limitations. As I was told on numerous occasions, it is better to embrace taxonomic disagreement than to wait for a professional taxonomic consensus that will never materialize. The extensibility of the taxonomic space, by way of filling gaps in the taxonomic record to increase the access-oriented powers of a consensus space, has the potential of building structures far more robust than they otherwise would be. This is especially important for external organizations that use the Catalogue as a core element of their data hierarchy. The external functions of the Catalogue are far more generative and provide the maximum amount of extensivity in the biodiversity infrastructure environment.

EXTERNAL EXTENSIBILITY

While the internal extension of the Catalogue increases the breadth of taxon coverage within the system, the external capacities of these backbones radiate outward far more impressively. An essential access-oriented aspect of consensus structures is that they serve as a modeling mechanism for data all over the world. One might think of this kind of use as *taxonomic amplification*, defined as the use of a composite taxonomy to collocate disparate data from multiple sources and to produce taxonomic knowledge by the appropriation and recombination of a taxonomic structure, harkening back to Gerald Guala's concept of synonymic amplification (2016). The Catalogue is intended to support the shift from local practices (hypothesis driven, internally consistent) to global spaces that rely on the embedded standardization for full-scale integration within global infrastructure to support "a quantum increase in the coherence of the world's biodiversity data and analyses" (Species 2000 2015a). And this use of classifications—and the way these systems change because of it—is vitally important to understand.

Derek Langridge (1992) described some of the roles and applications of classifications, including facilitating searching and arranging documents. Extensibility is far more invasive in terms of how it intersects with the user-interface environment, mostly because taxonomies are the foundation for all data management. Langridge makes the point that computers do not eliminate the necessity for classification, they merely provide a more sophisticated and effective means to create more nuanced (and obfuscated) classifications (1992, 70). Indeed, this statement still holds true, for the importance of classification, especially with regard to databases and computational modeling environments, continues to rise even more than twenty-five years after Langridge's publication. If databases have done anything, they have forced us to represent and classify *everything* to the minutest detail so as to facilitate the transfer of knowledge within computational spaces. The existence and subsequent widespread use of biodiversity consensus structures is a testament to such a fact, as is the extent to which they structure the understanding and representation of global knowledge. If, as Jerome McGann (2001, xi) states, "our minds think in textual codes," we might also extend this statement to declare that the public and social conception of biodiversity knowledge "thinks in taxonomic codes" through the lenses of these networked structures. And, more often than not, these taxonomic codes are consensus-based. If this is the case—and my contention is that it is increasingly becoming the standard—we need to think more about what this means for the agents and organizations that put these standards to use, and how we might theorize the kinds of changes that are taking place as part of this integration and rearticulation.

A good place to start this line of inquiry is with Joseph Tennis (2015), who postulates the existence of three essential theoretical approaches to the study of classification that define the kinds of work being performed in this domain. Tennis's categories are foundational classification theory, first-order classification theory, and second-order classification theory. First, foundational classification theory, as defined by Tennis, is "concerned with philosophical and definitional aspects of classification" (2015, 246). A. Broadfield's (1946) work is an example of such an approach, as would be the categorical and philosophical expositions of Bliss, Langridge, Richardson, and Wilson. Larger questions in this arena include ontological and epistemological questions about the a priori assumptions structuring organizing systems. One might ask, as part of this level, What are the epistemic differences between traditional and consensus biodiversity structures?

First-order classification, Tennis's second category, is "solely concerned with the methods of classification scheme construction and use" (Tennis 2015, 245). This definition helps us understand the nuts and bolts of system construction—how we build classifications and articulate the processes we manufacture to produce them. Previous chapters expanded on such first-order and also foundational approaches: how the Catalogue conceives of and constructs evidence for concepts, how such concepts can be connected within a historicized nomenclatural system, and how management classifications build their schematics in contradistinction to traditional taxonomic forms.

Third, and most pertinent to our current discussion, is second-order classification theory, which is concerned with the use, reuse, and manipulation of systems once they have been completed (Tennis 2015, 246). With regard to the Catalogue, we are interested in how its management hierarchy is used in various contexts and how that intended use fundamentally changes the Catalogue's composition—and impacts the production and representation of knowledge as it becomes a vehicle for data in both anticipated and unanticipated spaces. Second-order classification is then broken down into three subcategories: (1) how schemes change over time and how we update them, (2) how installed schemes interoperate, and (3) how systems change when they "change context (reapplied or reengineered)" (Tennis 2015, 246). Each of these elements is pertinent to biodiversity databases, but the third is most relevant in relation to the extensive qualities of the Catalogue.

Understanding how the Catalogue is repurposed, reengineered, and changed is absolutely critical if we are to apprehend the impact of the Catalogue on the shape of public knowledge. Modifying a classification enhances the system in some attributes and limits it in other ways (Tennis 2015, 246). Many of the database systems in the biodiversity world, for example, implement the Catalogue's taxonomy to some extent or another in order to organize and present its content. Does this matter? My contention is that it should, for anyone who works in database systems and classification. In the case of the Encyclopedia of Life, the Catalogue's consensus taxonomy is displayed as one option among an array of taxonomic approaches, alongside taxonomies contributed by National Center for Biotechnology Information, Barcode of Life Data Systems, and Wikipedia. GBIF's case is particularly interesting, given that their taxonomic backbone builds on the Catalogue's hierarchy to organize all the data compiled from sources that may or may not fit within any given consensus hierarchy. Let's expand on the case of GBIF.

The GBIF Backbone Taxonomy builds on the Catalogue's management hierarchy to organize all data uploaded into its management system. GBIF collects a vast array of data types, including occurrence records and nomenclatural data, as well as genetic data from sources such as the NCIB. This is a functionally different emphasis than that of the Catalogue. The Catalogue's primary function is to provide valid and accepted name tokens (as a nomenclature) and the subsequent connection of these tokens into valid taxon groups. GBIF's main function, on the other hand, is to collect globally produced occurrence data and provide a point of access for data points that can potentially differ in scale and that may or may not point to an accepted species name (GBIF 2020). Validated, curated names and taxonomic relationships are the primary focus for the Catalogue, whereas in the GBIF environment, the occurrence data takes priority, which is appended to the best-fit taxon level to aid in data access and use. This isn't to say that GBIF isn't concerned with validated names and most-current classifications (it, of course, ultimately sees this as a vitally important part of the process); this is just to say that the main emphases of each organization influence how they approach the process of building classifications and what concessions they are willing to make in the process of doing so. The Catalogue initially pushed against the use of algorithmic methods, for example, whereas GBIF embraced them–and to generally positive ends as conveyed by many biodiversity scientists. These approaches have changed in recent years, as collaboration between GBIF and the Catalogue has increased—the Catalogue of Life Plus emerged from a partnership between the two.

The Catalogue represents a significant percentage of the name and taxonomic information for GBIF's Backbone Taxonomy, comprising some 3,175,925 names, or approximately 54 percent of the total GBIF namespace ("GBIF Backbone Taxonomy—Constituents" 2017). However, even with the Catalogue's impressive 2 million species, the information falls short for GBIF's purposes (GBIF.org 2016). As a GBIF staff member conveyed, no global taxonomy can organize the more than 750-million-plus occurrence records that exist within the system. Some data is submitted with names not yet validated, and other data is formulated as the output of genetic analysis that may or may not be yet described at the species level. Information of this nature may or may not find a location in an inherited hierarchy, which means that GBIF needs to articulate methods to bridge these systemic differences. And so, they build on top of the Catalogue management hierarchy. In addition to the Catalogue, other trustworthy checklists are collected, including, for example, the International Plant Names Index (IPNI) and Index Fungorum (2021), to make an even more encompassing taxonomic whole. To accomplish this, GBIF has produced fine-tuned algorithms to offset the labor of this kind of work, which overlay hierarchies over hierarchies that are perfected across time and freely available on GitHub (Robertson 2016; GBIF 2017a, 2017b). And because GBIF gets a high degree of use, feedback is ongoing and iterated into new versions of the algorithms code (Robertson 2016).

Of course, this says nothing of the critiques one might levy against algorithms as organizing mechanisms, which Safiya Noble has written about extensively in recent years (2018). Many scientists feel that such methods are not as refined as taxonomies assembled through individual mediation, in addition to the more political concerns regarding the substitution of computational methods for hard-earned taxonomic expertise. Taxonomy is seen in many circles as a dwindling career, in part because of computational methods. The extent to which these taxonomy-building algorithms can remain effective is ultimately dependent on the quality of individual data sources. As Roderic Page, former science director for GBIF and professor of taxonomy at Glasgow University, has asserted on Twitter, "some data [GBIF does] have is poor (e.g., @catalogueoflife has mangled butterfly names)" (2016c). In addition, many of the data sources are "[aggregations] of sources that [may] themselves be aggregations," making it incredibly difficult to fix the data at the source (Page 2016a, 2016d). Minute editorial miscalculations inherent in the Catalogue of Life have to be debugged and restructured downstream to meet various infrastructure-specific needs.

Within this collocation activity, information conflicts inevitably emerge. The hundreds of millions of circulating records that find their way into GBIF collectively constitute an environment wherein species concepts, data types, data granularity, and taxonomic hierarchy formats may be in mismatch with one another, both semantically and syntactically. Data quality varies across the platform. Certain professional scientists and biodiversity informaticians, including Roderic Page, have indicated that perhaps GBIF should "take more 'ownership' of data quality, but that's politically tricky" (Page 2016b). True enough, indeed, as all data mediators should. But then again, one needs to weigh the benefit of access-oriented systems against that of those that strive for curatorial perfection. After all, to practice science and produce valid results, one needs data as a starting point. All taxonomic work is iterative, so consensus structures are certainly not immune to this fact.

EPISTEMIC INFLUENCE

What can we make of the Catalogue's impact on the organizing of global data? Even if the GBIF's nub taxonomy comprises many independent systems, given its overwhelming presence in GBIF, it is undeniable that the Catalogue impresses quite an impact on the constitution of knowledge in this space. Regardless of whether the Catalogue is a visible aspect of a system such as GBIF, it behooves us to unpack how the assumptions that underlie its construction deeply influence the shape of knowledge throughout the biotaxonomy environment and beyond. Web-based media are rife with representational and graphical structures, and biodiversity database expressions are particularly laden with mapping software that expresses data in easily consumable formulations. GBIF is powerful in this respect—the opening webpage displays a marvelous map-based analysis of its data that incudes data density by geography, year, and the basis of a record (observation, literature, preserved specimen, and so on) (GBIF 2017c). Yet, as Johanna Drucker (2014b) reminds us, the space between data and the graphical interface is filled with interpretive acts—acts that begin with the initial construction of data structures that organize how knowledge should and can be displayed for ready access. One of the primary goals of unpacking and theorizing structures like the Catalogue is to narrow this interpretive gap between data structures and the mediated mechanisms by which we access those data in precoordinated graphical environments.

It goes without saying that the usefulness of these sites as tools that facilitate biodiversity work is powerful to certain extents, so the purpose of such a critique is merely to shed light on the machinations that make *all* taxonomies useful, to highlight how our results might be weighted in favor of certain epistemological values—and, in this particular space, the Catalogue stands as a central mechanism by which those values are constructed. This kind of deconstructive taxonomic work is particularly important given the extent to which sites like the Encyclopedia of Life are implemented in educational environments, illustrated by the integration of educational resources aimed at a primary-school level audience. But given the focus of other biodiversity data platforms such as GBIF and the International Barcode of Life, the impact of this consensus work extends beyond the classroom, affecting scientific research in a wide array of subdomains. GBIF has attained such widespread use that the question of how it is made is often overlooked, as data points are downloaded, referenced, and integrated into the practice of analysis, interpolation, and interpretation.

As a case in point, I had a meeting with an evolutionary biologist studying the sleep patterns of biological organisms. Their team was using GBIF data to fill in some ancillary data elements that weren't part of their original collected data sets. In this case, the research group had collected primarily individual, organism-level information (sleep patterns for one organism, for example), and such individual records needed to be placed alongside other species records to create a kind of ad hoc taxonomy depicting sleeping behavior. To create this map, species-level data was needed, to connect individual specimen data with larger species range and distribution trends. This data also helped scientists map their data using GIS software. And although this particular individual knew that taxonomic errors were possible in the case of any platform—and indeed, in this case, data was confirmed at the data source to confirm its validity—they were intrigued to learn about the various layers of aggregation, consensus work, and interpolation that were working together to create the robust knowledge structure represented by the Catalogue, and, to a certain extent GBIF. Taxonomies become invisible to users, so making them visible helps one understand their capacities—and, more importantly their limitations—in structuring easy and ethical delivery of information. An undergraduate biology student just beginning a research career would likely not have been as savvy as these scientists—which says nothing of the lack of expertise a politician or journalist might bring to the table.

The materiality of consensus taxonomies, then, becomes a central point of examination as we think about consensus structures and their impact on the production of new forms and domains of knowledge. In Karin Knorr Cetina's work, especially *Epistemic Cultures*, she offers a depiction of scientific knowledge societies that are fundamentally tied "to the machines deployed in knowledge production" (1999, 11) and notes that the scientists who perform the process of knowledge-creation are derivative agents to the instruments and tools that structure and facilitate their examinations. In Cetina's analysis, instruments limit the capacity of what can be done within a laboratory setting, but perhaps more importantly, these central objects create a bounded, ontological reality that communities become embedded within as part of daily practice. How these instruments are deployed both rhetorically and materially has fundamental impacts on how the cultures that employ them arrange themselves—consciously or not—but also on how they construct a narrative for the production of scientific activities and analysis. Scientists relate to machines quite differently in the domain of high energy physics, for example, than within the space of molecular biology. In the former, instruments take on a symbolic quality and become the internalized, closed, and primary ontological system of the lab against which scientists define their intervention. In the latter, however, the focus is not on a symbolized system, but rather on analysis of the external natural world—mice specimens, for example—that scientists then control toward the goal of knowledge through experimentation. The "real" of a laboratory (the space of primary intellectual focus and interest) is created, in part, from how instruments embody and build a narrative that constructs the way an investigation can take place.

My goal in bringing up Cetina's work is to illustrate that systems (tools, machines) like the Catalogue have the capacity to change the nature of particular scientific work based on the suppositions these systems make about the "real" objects they purport to organize. The existence of consensus structures allows scientists the freedom to perform new kinds of practices based on an entirely different set of expectations about what these taxonomies are trying to say about the natural world. In the case of description-based taxonomies, the "real" biological world is presumed to be mimicked (perhaps idiosyncratically, and certainly hypothetically) within a closed system that functions under a uniform set of suppositions about how to define a taxon and how that taxon is related to another based on a series of metrics. The "real" in the case of retrieval-oriented consensus spaces is markedly different, and far more speculative and processual. Consensus structures mimic an amalgam of opinions about how we can potentially classify the natural world through the process of mediation. The focus becomes not a reflective, empirical model of the natural world, but the processes, practices, and struggles for instantiative power among many competing taxonomic spaces. And so, as individuals, systems, and organizations integrate consensus data into their workflows, these underlying suppositions must be, first, acknowledged, and second, negotiated and rendered transparent through the access systems they employ.

With this new kind of consensus-based epistemic influence, new forms of knowledge can arise that build on these suppositions, most notably new initiatives focused on using repositories like the Catalogue for predictive purposes.

HORIZONS

A more experimental, but nonetheless powerful, extensive ability of consensus systems is their potential use as a springboard from which to assess larger trends in the taxonomic world. Nomenclatural networks, designed as they are to trace the historical connection and development of concepts via a system of internal taxonomic relationships, provide a rich space of knowledge about our ever-changing understanding of the natural world. As systems like the Catalogue are integrated into backbone systems such as those created by GBIF (which is a proper consensus system in its own right), these consensus structures organize a huge variety of information about species, including geographic distribution, migration patterns, ecological and species relationships, genetic information, and more. Collectively, these data have the potential to help scientists—and especially evolutionary biologists—understand historical trends regarding how scientists have, historically, documented and organized biodiversity. And such historicity can potentially help predict some of these attributes.

As a result, these repositories need to be able to capture the full scale of taxonomic opinion over time to appropriately model evolutionary patterns in a multitude of ways. As taxonomist Nico Franz has noted, "The real-life challenge for these information repositories is to capture more than one authoritative classification; they are built to represent the full spatial and temporal dynamic of the taxonomic process" (2005, 499). Capturing this dynamism is essential, especially since taxonomic production (and the interpretation of the species in the natural world), as a scientific practice, is constantly changing. The infrastructures that we build to support and document these processes should be able to mirror the various transmutations of taxonomic opinion and representation that happen over time.

Building on this, a key extensive capacity of consensus systems is to provide the capabilities to extrapolate "macropatterns" (Bourgoin 2016) using the documentation of historical concepts within the database structure, as well as the associated data that is mapped to these species concepts. In figure 6.2, Thierry Bourgoin, faculty member at the Muséum National d'Histoire Naturelle in Paris, illustrates a possible schematic for how we can think about knowledge bases as they relate to structures such as the Catalogue. Bourgoin makes a distinction among classification, phylogeny, and evolution, as discrete layers needed for the production of knowledge bases. Within this schema, the production of a classification is a preliminary step, which depends on the network of name associations lying at the core of most biodiversity platforms. These names are then synthesized into

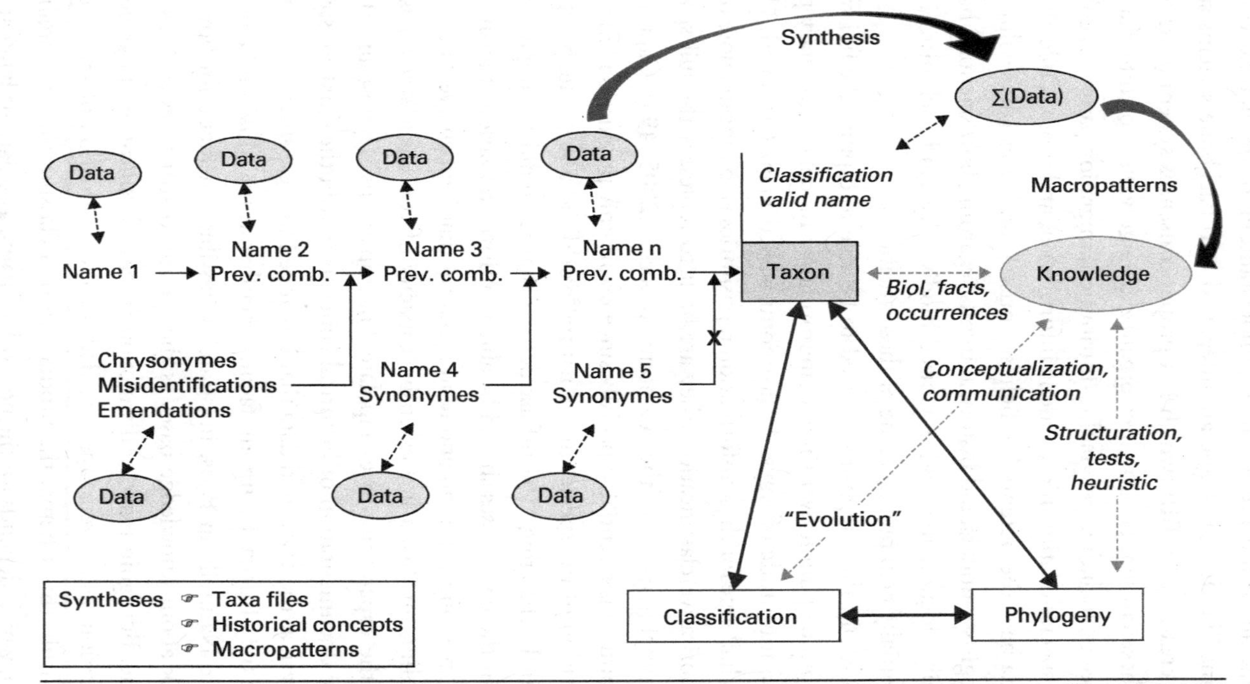

Figure 6.2
From taxonomic databases to knowledge bases: understanding evolution. The goal in a system of this nature is to (a) take name data that is synthesized into validated taxon forms and (b) use these taxa as the building blocks for classifications that can then be used to articulate many possible phylogenetic hypotheses.
Source: Thierry Bourgoin (Bourgoin 2016). Used by permission.

accepted taxon groups, which can then be iteratively placed in any number of classificatory arrangements. The ultimate, long-view goal is for platforms such as the Catalogue to confront phylogenetic and ecological questions so as to better "understand evolutionary [concerns]" (Bourgoin 2016). Such a broad and extensive approach to speculative work of this nature, however, requires a networked set of coordinated infrastructures, each with their specific data contribution to this broader whole.

Biologists and informaticians have termed this kind of macro-level analysis, among other monikers, *evolutionary informatics*, which "concerns the capturing, storing and integrating of all these data (about biological specimens, images, genomes, etc.), as well as developing the analytical techniques that use them to answer evolutionary questions" (Parr et al. 2012, 94–95). Platforms such as the Catalogue of Life stand at the center (Parr et al. 2012, 100) of this informatics integration, either as part of the taxonomic backbone to platforms like GBIF or as a source of valid names and synonyms by which all the data within a system is collocated. This makes the Catalogue a key, but certainly not the only, player in the constitution of any knowledge base. But the biases engrained in the Catalogue's construction must be negotiated and acknowledged in these integrative environments. As Parr et al. (2012), indicate, an endeavor such as this collocates many discrete data types, standards, and integrated repositories in a way that allows them to function as a unified whole—but also in a way that has not yet been perfected in biodiversity work. Aside from those organizations that are part of the iLife consortium, the authors identify a broad number of platforms that must integrate for such predictive functionality, including the Map of Life (2018), which provides geographic distributions; GenBank, a repository for gene sequences; and MorphBank, which offers image annotation functionality. This entire ecology of historical documentation provides a sandbox of potential and emergent knowledge products beyond the sum of the database content itself. Such work recombines data in ways that can provide long-term predictions, climate and species modeling, and simulation (Landers 2016).

Large-scale coordination of this nature takes many years to articulate and perfect. For starters, the success of this scale of data integration will involve the careful documentation of how taxon concepts, and the taxonomies they are embedded in, change over time. Capturing this contingency in some useable, structured, and representationally sound manner is of utmost importance. It has been acknowledged that the long-term prospects and success of the Catalogue depends, in one sense, on the system's transformation from one that organizes a stable taxonomic backbone that facilitates data sharing to a system that can map the changes that are taking place over time within its own database, as well as among the databases it inherits as part of its backbone taxonomy. The complications involved with the historical tracing of this kind of change, however, far exceeds the capacity of the Catalogue alone. The successful prediction of evolutionary trends will rely on the successful evolution of biodiversity practices, in which coordination and collaboration are central tenets of their mission and values. This is precisely why consensus structures are such important elements in biodiversity work today: they push the practice of taxonomic work beyond its traditional epistemological boundaries, and in doing so, support the articulation of new generative spaces of scientific speculation.

Trying to accommodate a complex arrangement of multiple taxonomies, as well as document how those arrangements evolve over time, is no easy task, particularly when such complexity must somehow be fixed in database fields that are relatively limited in their composition. Implementing knowledge bases is, in many ways, pushing the boundaries of what consensus spaces are situated to perform. As Johanna Drucker conveys, "knowledge forms are never stable or self-identical but always situated within conditions of use" (2014b, xiv). If we embrace taxonomy as an iterative process, then the platforms and structures that we use to collocate this knowledge should be seen as in-process. Consensus spaces are tied to the beliefs of a particular space and time—as are any organizing structures. Indeed, even the predictive knowledge produced from these systems is speculative and hypothesis-oriented, not gospel. But neither is any descriptive taxonomic system a final opinion. The motivational core of consensus systems is not that they strive for perfection or epistemic constancy, but that their imperfections are to be completely on display, so that the users of these systems can use data responsibly and with a keen understanding of what knowledge can be produced from these platforms.

It would be a mistake, I think, to assume that documentary classifications such as the DDC do not also radiate their epistemic and structural capacities in spaces far beyond their boundaries in ways that are not dissimilar to the Catalogue. The previous discussion on scientific warrant unearthed some of these effects—library systems also are built with particular values and have material consequences. Thinking about how taxonomic knowledge is being enacted, negotiated, and compromised within consensus biodiversity work can perhaps help us reformulate the very foundation of what we mean by *control* in documentary classificatory spaces within information science. Should we rethink the potential of instrumental powers in light of lessons learned from the Catalogue? Perhaps our systems need to provide the flexibility to design organizing systems not only for consistency and stability, but also that are understood as emergent, contingent knowledge in their own right. The *epistemology of contingency* has broadened the horizons of scientific taxonomic work—a domain with roots far deeper in history, and more rigidly theorized, than documentary classification in information science. How might we in IS reimagine our systems in light of these developments? Would such an approach increase diversity of opinion within taxonomic spaces? For now, however, let me table this argument, and turn our attention to a critique of consensus systems, which will place us in a much more effective position to judge their possible efficacy for documentary classifications.

The domain of information science has seen a resurgence of literature that critiques organizing systems as biased, culturally situated, and potentially suppressive and oppressive infrastructures (Olson 2002; Adler 2017; Noble 2018; Furner 2009a; Bowker and Star 1999). These critiques have been useful toward conceptualizing how information science can construct classifications that are more robust and attentive to the needs, sensitivities, and realities of a diverse set of users. It is only fitting then, now that we've talked about the virtues of consensus systems, that we assess the limits of these hierarchies. It is certainly true that consensus systems are not universally accepted in practice. In fact, some professionals feel they do more harm than good to the biodiversity taxonomic world, undermining the processes, assumptions, and use of description-oriented systems, which are crafted though hard scientific work and empirical analysis. For the field of information and library science, the implications for these critiques are broad and deep. Critiques levied toward scientific consensus systems can help information scholars identify similar weaknesses in documentary-based consensus systems. Secondarily, these implications can help rearticulate information systems that are specifically built for flexibility and that strive to manage consensus while remaining attuned to the multiple interpretative layers built into their design.

7 EPISTEMIC CONFLICT

CONFLICTS

For all the benefits that composite classifications offer the data-rich biodiversity world, such an approach has not been embraced full-scale by all taxonomists and bioinformaticians. And this is for good reason, both technical and intellectual. Stephen Thorpe's general critique of the management classification approach on the Taxacom listserv can set the stage for these critiques:

> Biological classification is a mixture of scientific fact . . . and subjective opinion. . . . Both these factors taken together doesn't make life very easy, and it is all in perpetual flux. . . . However, I don't think that the issue can be "managed" in quite the way that is envisaged by some. . . . My primary governing principle is that, subject to monophyly, classification is primarily a filing system to make information management easier. So, it doesn't really matter which classification is followed, PROVIDING that it is explicitly stated which one. The problem with adopting a particular classification for a large group (like the "Protista") is that advances in taxonomy happen on much smaller subgroups, so if you blindly follow one particular broad classification, then you cannot accommodate the advances very easily. Hence, I think you have to simply treat matters on a case-by-case basis, and just choose and specify a sensible classification for that particular case (and change it, if necessary, if something more convincing is published). To try to come up with a single "officially endorsed" classification would simply be to ignore the subjectivity and fallibility of taxonomy. (2009)

In general, critiques come from two directions. On the one hand, a composite classification presents technical complications, including issues such as proliferating data error, assessing data quality, and the like. These technical concerns, while vital to the integrity of the system, are not altogether new in the realm of database work. Thorpe's view certainly seems reasonable in this respect, especially given the fact that such top-down approaches to classification are antithetical to the way taxonomy has functioned for hundreds of years (Godfray 2002). Contemporary biodiversity practice, however, does need to attend to the ever-increasing rates of data production, and large-scale questions about environmental issues and mass extinctions are driving the need for these more-integrated consensus approaches (Guralnick and Hill 2009). Additionally, maintaining numerous independent taxonomic systems is not necessarily a pragmatic approach. This said, while the world checklist system and consensus-oriented view might work at the production level, "for advanced uses, it's sh*t," as one prominent taxonomist poetically proclaimed during one of our conversations. Finding the balance between generalized and expert system design is easier said than performed in practice.

The more serious critiques, at least as they relate to the main narrative of this book, concern the epistemic conflicts that occur, since, when multiple taxonomies are integrated into a composite space, they lose nuance and are manipulated in ways that contradict local scientific practice. The reality is that contributed taxonomies are "qualitatively transformed" as they are ingested into the compiled system (Remsen 2010). As databases enter the Catalogue of Life, the editors do their best to maintain the integrity of the original taxonomic structure, but there are times when "adjustments may need to be decided upon by the editors on where and how to insert it, to make it as consistent as possible [with the rest of the Catalogue of Life], while not losing the essential taxonomic information it has been created to provide" (Species 2000 2016a). Let us assume for one moment that we are contributing a taxonomy to the Catalogue. And let us further assume that conflicts arise between our contributed piece and the Catalogue's existing structure. One can make an argument that, however trained and expert an editorial board might be, the best person to make a determination about the integrity of the contributed taxonomy is the original creator. Losing their intellectual agency and authority is of great concern to contributing taxonomists.

To concretize how this conflict might transpire, refer to figures 7.1 and 7.2, which compare the taxonomic hierarchy from both the combined Species 2000 and ITIS Catalogue of Life (figure 7.1) and the stand-alone Integrated Taxonomic Information System (figure 7.2), to see how this kind of editing might take place within database environments. The ITIS taxonomic hierarchy is far more detailed in its composition, particularly because it contains a more finely articulated higher-level taxonomic ranking system (inclusive of subkingdom, subphylum, infraphylum, superorder, etc.). While some might say that this information isn't vital to a species-level articulation, other might note that the existence of higher-level taxa is important for contextual reasons. In comparison, looking at the Catalogue of Life's taxonomic tree, you can see that the Catalogue editors chose the

Recognized by	Rank	Classification
• **Species 2000 & ITIS Catalogue of Life: April 2013** View in classification	Species	Animalia + Chordata + Mammalia + Carnivora + Ursidae + Ursus + ***Ursus arctos* Linnaeus, 1758** *Ursus arctos alascensis* Merriam, 1896 *Ursus arctos arctos* Linnaeus, 1758 *Ursus arctos beringianus* Middendorff, 1851 *Ursus arctos californicus* Merriam, 1896 *Ursus arctos collaris* F. G. Cuvier, 1824 *Ursus arctos crowtheri* Schinz, 1844 *Ursus arctos dalli* Merriam, 1896 *Ursus arctos gyas* Merriam, 1902 *Ursus arctos horribilis* Ord, 1815 *Ursus arctos isabellinus* Horsfield, 1826 6 more... show full tree... *Ursus americanus* Pallas, 1780 + *Ursus maritimus* Phipps, 1774 *Ursus thibetanus* G. [Baron] Cuvier, 1823 +

Figure 7.1
Two different curated taxonomies displayed by the Encyclopedia of Life for the species *Ursus arctos* (Encyclopedia of Life 2017). CC-BY 4.0, Catalog of Life, used by permission.

Integrated Taxonomic Information System (ITIS)
View in classification

Species
- Animalia +
  - Bilateria +
    - Deuterostomia +
      - Chordata +
        - Vertebrata +
          - Gnathostomata +
            - Tetrapoda +
              - Mammalia Linnaeus, 1758 +
                - Theria Parker and Haswell, 1897 +
                  - Eutheria Gill, 1872 +
                    - Carnivora Bowdich, 1821 +
                      - Caniformia Kretzoi, 1938 +
                        - Ursidae Fischer de Waldheim, 1817 +
                          - Ursus Linnaeus, 1758 +
                            - ***Ursus arctos*** **Linnaeus, 1758**
                              - *Ursus arctos alascensis* Merriam, 1896
                              - *Ursus arctos arctos* Linnaeus, 1758
                              - *Ursus arctos beringianus* Middendorff, 1851
                              - *Ursus arctos californicus* Merriam, 1896
                              - *Ursus arctos collaris* F. G. Cuvier, 1824
                              - *Ursus arctos crowtheri* Schinz, 1844
                              - *Ursus arctos dalli* Merriam, 1896
                              - *Ursus arctos gyas* Merriam, 1902
                              - *Ursus arctos horribilis* Ord, 1815
                              - *Ursus arctos isabellinus* Horsfield, 1826
                              - 6 more... show full tree...
                            - *Ursus americanus* Pallas, 1780 +
                            - *Ursus maritimus* Phipps, 1774
                            - *Ursus thibetanus* G. [Baron] Cuvier, 1823 +

Figure 7.2
The (default) classification hierarchy for the species provided by the Catalogue of Life. (Bottom) The classification hierarchy provided by ITIS. CC-BY 4.0, Catalog of Life, used by permission.

genus node as the connection point for the ITIS database. Note the lack of an author and publication designation after the species epithet, *Ursus*, at the genus level in the Species 2000 taxonomy. Additionally, the Species 2000/Catalogue of Life hierarchy retains the ITIS taxa detail below the genus node connection point (beginning at *Ursus arctos* Linnaeus 1758), whereas above that connection point, the Catalogue's management classification backbone maintains authority. The act of choosing one part of the ITIS classification over any other transforms the ITIS structure in fundamental ways, divorcing *Ursus arctos* from the upper-level backbone context in its original location in the ITIS database. In all, this example is fairly generic, but other decisions can be far more detrimental.

And so are the dangers of universal and consensus classification systems: they normalize data in ways that obfuscate different epistemic points of view. Sterner, Witteveen, and Franz address this fact in their article, "Coordinating Dissent as an Alternative to Consensus Classification: Insights from Systematics for Bio-ontologies." The authors identify three types of consensus systems that have "advantages in certain contexts and for certain ends" (2020, 2). The first is the Definitional Consensus Principle (DCP), a classification system intended to be internally coherent (in the sense Jonathan Furner notes) and one that represents consensus about the definitions of entities included in the classification. DCP is further bifurcated into two types: (1) one that adheres to ontological realism as articulated by Barry Smith and Werner Ceusters (2010) and is termed Realist Interpretation (DCP/R), and (2) one that is based on local context, termed a Contextual Interpretation (DCP/C). DCP/R maintains that to limit the production of multiple competing biological (computational) ontologies, centralized reference systems should be created that represent "settled science" (Barry Smith and Ceusters 2010, 1) as it expresses some true external reality (realist). Such a metaphysical approach would presume a uniform (philosophical) ontological understanding of universals, types, and properties (inclusive of relations) that should then be formulated into bio-computational ontologies. DCP/C, on the other hand, attends to classificatory consensus at a local level that represents that particular group's (and potentially *only* that group's) metaphysical or epistemic principles that, might, "for example, characterize what counts as mature or settled science" in that context (Sterner, Witteveen, and Franz 2020, 10).

In response to DCP positions, Sterner, Witteveen and Franz (2020, 7–20) indicate a consensus system should focus not on the ontological or epistemic aspects of classification (that is, to produce a taxonomy that is presumed to be *correct* in both of these senses), but rather on aspects meant to coordinate names, primarily, with the underlying taxonomic structure acknowledged as a mechanism for communication. In some ways, this approach is aligned with that of the Catalogue of Life or the GBIF Backbone taxonomy, with the exception that systems such as the Catalogue are often touted as representing, at least at the sub-consensus level, a "correct" or "current" taxonomic arrangement. A coordinated consensus principle (CCP) approach eschews this rhetoric of correctness entirely and claims only that the names and the type it refers to—collectively known as the taxonomic concept—are valid in this space. Such a system should make no claims as to how these taxon concepts might then relate to the articulation of a species concept or larger taxonomic ecology—this interpretive work should be left to the local and contextual classification, as they always have been.

Adding to the complexity of these epistemic conflicts is that taxonomic opinions are not only in a constant state of change but they also do so quite frequently. This taxonomic contingency will perpetually be in tension with the urge to control change in "officially endorsed" systems designed for stability such as the Catalogue. Such ontogenic concerns have been of increasing interest in the information studies community (Tennis 2002, 2012, 2015), concerned as we are with how to theorize and manage the lateral transformations of classification systems. Such change is important if we are to keep them relevant to our organizational and access concerns. In the biodiversity world, change becomes an ever-more-present issue as consensus structures have emerged on the scene. For all their virtues, consensus structures have not been designed to track internal structural changes. And in biodiversity space, structural change is always produced by a shift or update to some theoretical condition.

As Tennis notes, ontogeny is the examination of "the life of a subject over time—the subject's scheme history" (2012, 1351). Ontogeny, or ontogenesis, is a surprisingly appropriate term for the development of classifications in our context, given its general usage in the field of biology to describe the mechanism and development of individual organisms. Ernst Haeckel, in *Generelle Morphologie der Organismen* (1866), defines four general concepts: ontologie, phylogenie, ontogenesis, phylogenese. Løvtrup summarizes Haeckel's concepts in this way: "'Ontogenie': the history of the development of the individuals. 'Phylogenie': the history of the development (evolution) of the taxa." Ontogenie refers to the development of individual organisms—as in the progressive and regressive development of one organism over its life span from its embryonic stage to death (Løvtrup 1987, 201). Phylogenie, on the other hand, refers to the historical reconstruction of taxa as represented in phylogenetic classifications. In the context of biodiversity classification, it is imperative that we are able to trace both the changes that taxon concepts undergo and the changes that might take place with regard to how these are activated within classification systems and related as species with various hierarchies.

Tracing this change over time is incredibly difficult in the space of a composite classification, particularly because change is happening at two levels, both of which are nearly impossible to track. On one hand, we have the change that is occurring within local taxonomies that are contributed to the backbone. On the other hand, we have the changes that are occurring by way of editorial intervention in the Catalogue itself. These are both crucial pieces of information for taxonomic experts, and one of the primary reasons that consensus systems often fall short of their intended use. As we've discussed, contributed taxonomies are snipped out of their original context for inclusion into the Catalogue, but the minute editorial changes that are taking place are difficult, if not impossible, to reconstruct. For experts, working with such a moving target is not a tenable situation. The position of a species is also continually redefined in the Catalogue's space, if not by its absolute position, then by its relative position. As taxonomies are incorporated into the Catalogue, they are set alongside other contributed taxonomies. So, by virtue of these juxtapositions, the relative position of one subject to another is constantly transformed. Understanding these relative relationships is essential to understanding the internal logic of any given taxonomy. In *The Organization of Knowledge and the System of the Sciences*, Bliss speaks of the order of disciplines in the Dewey Decimal Classification schedule and how the placement of topics may cause confusion: "Methodology (112) stands between Ontology and Cosmology and far from Logic, of which it is usually regarded as an extension" (1933, 218). Olson also points to the definitional qualities of classificatory structures: "The hierarchy thus created structured knowledge by putting every subject in its place. It creates a context for each subject within this hierarchical arrangement" (2002, 22). Within biological taxonomies, a species' relative position with regard to other species is of great consequence, both theoretically and methodologically. To change a position is to change a species' overall context.

EXPERTISE, NUANCE, AND MAINTENANCE

Another pressing issue of concern is that these composite classifications extend beyond the boundaries of the scientific domain itself, meaning that their impact beyond their intended audience must be considered. This is a major point of friction within the biodiversity community. Increasing numbers of scientists and historians of science worry that such taxonomic backbones, and the corresponding epistemic and ontological assumptions they convey, undermine the authority of taxonomic opinion more broadly speaking. If we return to our case of the classification of the Australian dingo, we can see how the use of these systems by nonspecialists would present some serious material problems. On the face of it, the Catalogue inherits the classification of the dingo as represented in ITIS, where the dingo is categorized as *Canis lupus dingo*, a bona fide species of its own. However, there is nothing to say that some other consensus classification might (perhaps wrongly or correctly) classify the dingo as a feral wild dog, under the species *Canis familiaris dingo*. Under the circumstances of the latter, policy makers might rule the dingo unworthy of protection and order the mass eradication of the species on the basis of it being a pest. The crux of this matter is that, while the specialist may be able to understand the nuances of the dingo's classification, in the space of a composite classification, there is no indication that this conflict exists within the taxonomic community to a nonspecialist user. And herein lies one of the dangers of the consensus system: if systems claim to be validated and managed by experts, the prevailing assumption is likely to be that the entire system is, in fact, accurate. And given that "accuracy" proper is not an appropriate way to view taxonomic hypotheses, since there are many viable ways of constructing a classification, this nuance is important.

A larger extension of this problem is how to guide users through the taxonomy in a way that they are able to navigate the space with a critical eye that can spot issues related to quality and completeness. Some of these problems are obvious and clearly present themselves as errors, whereas others are so nuanced that they are never questioned. This is one way that power is systemic—it is obfuscated behind technics. Classifications of this magnitude are assembled from many moving parts, and at any of these junctures, the introduction of error is not only a distinct possibility, it may be even likely. Given the size of the Catalogue, the introduction of error is not often the fault of any one person, but is rather a consequence of rote data manipulation or local taxonomic practices. Problems arise when users are not versed in understanding how systems function, or what they are and are not meant to express. As Franz and Sterner indicate, as errors proliferate in these consensus systems, trust in them dwindles (2018). "Error," however, ranges from small issues, such as name misspellings, to more serious issues such as species miscategorization.

Robert Mesibov (2018), a biochemist and active taxonomist in the realm of biodiversity informatics, conveys a peculiar and fascinating example that illustrates the proliferation of erroneous taxon records within database environments. During an audit of the GBIF database, Mesibov points to his discovery of the invalid species name, *Not Chan*, 2016, that led him to examine the extent of similar problems in taxonomic databases. The "Not problem" shows how markers of unknown identifications in databases (represented by terms such as, not found, not given, not known, not identified, not listed, and so on) find their way into aggregated spaces despite their invalid nature. Subsequent examination led Mesibov to find similar issues within the Catalogue and Encyclopedia of Life. The result in this case is that multiple and similar kinds of nomenclatural and concept error were collocated across the global data landscape. Mesibov asserts that consensus backbones such as those produced by the Catalogue and GBIF provide a space in which these errors can "come to light" (Page 2018)—errors that would have otherwise been retained at the local level. A global view affords a perspective that a fragmented data landscape cannot. But even still, the errors proliferate until these discrepancies are located and fixed—and more often than not, it is impossible to track down all the systems that have integrated this error into their infrastructures.

Certainly, one way we can assess quality is through the credibility and reputation of the institutions and individuals that provide information to the Catalogue. But even the most credible sources are human, and errors are a fact of life. Such credibility is made visible within systems by linking back to the contributing GSD or RSD, as well as requiring the "Latest Taxonomic Scrutiny" data field for all contributed data sets to the Catalogue. This field group must include, the "name(s) of the taxonomic expert or editor, who is responsible for the taxonomic concept accepted in the source database and (b) date when the expert or editor [or small team] assessed the record" (Species 2000 2014, 10). Such linkages and attribution are essential in deconstructing the veracity of the Catalogue's individual entries, especially since the taxonomic editors depend entirely on the expertise—and reputation—of those that contribute this data. Once again, however, the underlying assumption is that the lay user (a) will question the data *at all*, and (b) if they do, they are willing to perform the legwork of verifying a particular source. To the latter point, one need only be employed as a librarian for a few weeks to realize that almost nobody, in fact, verifies sources. Such is the reality in a world where misinformation is running rampant among users who take information at face value.

In addition to these provenance markers, source databases in the Catalogue provide a *confidence rating*, or data set qualifiers, for the taxonomic data, which is certainly a step in the right direction (Species 2000 2017b). This information is then displayed in the main record at the appropriate taxon level (see figure 7.3). Of course, a major issue is that the quality rating itself is provided by the contributing database, and not by independent peer review, so the quality rating is dependent on the contributor's willingness and ability to acknowledge their own database's strength and limitations.

Still other data aggregators, such as the Global Biodiversity Information Facility (GBIF), have attempted to articulate more quantitative and decisive mechanisms for quality assessment. Acknowledging these rampant issues, GBIF formed a task force to articulate a series of recommendations on data fitness and efficacy for large-scale modeling use (Anderson et al. 2016). One of the main recommendations by the GBIF Task Force centered on a user's inability to accurately differentiate between quality and nonquality data:

> GBIF.org should serve indicators of precision, quality, and uncertainty of data that can be calculated practically, and preferably "on the fly," as well as summaries and metrics of completeness of inventories, at scales and for regions

Acidimicrobium ferrooxidans Clark & Norris, 1996

Name	*Acidimicrobium ferrooxidans* Clark & Norris, 1996	
Checklist status	Accepted species	
Classification	**Unranked**	*Biota*
	Kingdom	Bacteria Cavalier-Smith, 2002
	Subkingdom	Posibacteria Cavalier-Smith, 2002
	Phylum	Actinobacteria Cavalier-Smith, 2002
	Class	Actinobacteria Cavalier-Smith, 2002
	Subclass	Acidimicrobidae Stackebrandt et al., 1997
	Order	Acidimicrobiales Stackebrandt et al., 1997
	Suborder	Acidimicrobineae Garrity & Holt, 2001
	Family	Acidimicrobiaceae Stackebrandt et al., 1997
	Genus	*Acidimicrobium* Clark & Norris, 1996
	Species	*Acidimicrobium ferrooxidans* Clark & Norris, 1996
Source dataset	ITIS: The Integrated Taxonomic Information System 100% ★★★★★	
Link to original resource	https://www.itis.gov/servlet/SingleRpt/SingleRpt?search_topic=TSN&search_value=958726	

Figure 7.3

Catalogue of Life taxon record for *Acidimicrobium ferrooxidans* Clark and Norris, 1996. Note the "Source database" field, indicating the ITIS Global source database, the database version date (Sept 2015), the percentage of completeness of the species list this entry is embedded within, and finally, the confidence rating for the quality of the taxonomic checklist (level 5 value). CC-BY 4.0, Catalog of Life, used by permission.

> defined by the user. The summaries should display maps and graphs of completeness by region, time-period and taxa. (Anderson et al. 2016, 2)

The task force specifically recommended clear fields to indicate error and uncertainty rates, as well as provide methods for users to visualize data sets to understand the larger contours of that data and "highlight possible inconsistencies and error" (Anderson et al. 2016, 4). A valiant goal, to be sure, but often very difficult to implement.

It is of great import that governments and other funding sources have acknowledged the data maintenance, retrieval, management, and conservation, as an integral part of biodiversity work. If we look to large organizations such as GBIF, the Catalogue, and the Distributed System of Scientific Collections (DiSSCo), the European Union has paved the way for a more sustainable data landscape. GBIF subsidiary partners (global nodes) receive funding from a variety of funders, including individual governments, the

National Science Foundation, the Smithsonian Institution, and multiple ministries of science, education, and the environment, and the like. The reality is that, as the locus of biodiversity data power shifts from local, description-oriented work (individual scientists, research teams, and the like) to aggregative access-oriented data infrastructures, there is a real danger that the expert systematist community could see problems funding their own individual systematic projects. Certainly, GBIF and other organizations understand the problem inherent in this imbalance, and the Alliance for Biodiversity Knowledge (ABK) (Hobern et al. 2019) is one mechanism that has arisen in response to the current competitive research funding model. One goal of the ABK is to limit duplicative efforts on a number of fronts. Rather than everyone competing for limited funds to create many management taxonomies, for example, agreed-upon organizations should be designated to handle certain portions of the data landscape. The ABK also provides avenues for direct communication and participation from the research community to guide the path forward for such informatics-based projects.

However, even with this cooperative agreement, some taxonomists worry that the pendulum has swung too far in the direction of access, to the detriment of individual description-based research endeavors. And while the shift seems relatively benign, with the advent of large-scale infrastructures, some note that prioritizing data precision, stability, and comprehensiveness, rather than the systematic work itself only amplifies the epistemic problems that outfits like the Catalogue perpetuate. In addition, as organizations such as GBIF rise in prominence, they are increasingly becoming the mediators of biodiversity work with national and international governmental agencies. The Catalogue and GBIF are placed in a situation where they must balance both the needs of funders supporting their large-scale retrieval-oriented agenda and the requirements of the taxonomic community, which needs assurance that their taxonomic work will maintain a certain level of integrity in these systems. A fine line to balance, to be sure—and one that will only amplify in the coming years as this trend continues. And while the need to balance these two constituents may not be a bad thing, the reality is that such a position is indicative of the drastic ways power has shifted from the lab, so to speak, to the global management of accumulated data. As is often colloquially stated, if you want to know where the power is, follow the streams of money.

Divergent Traditions and Nameless Taxa

Perhaps the greatest technical issue for composite spaces is how to manage taxa tokens (names or codes) that do not conform to Linnaean binomial formulations. Genetic markers in the form of DNA barcodes (such as the mitochondrial C01 gene sequence) have been increasingly useful in constructing phylogenies (Waterton, Ellis, and Wynne 2013; Erickson and Driskell 2012). One result of the increasingly popular approach of phylogenetic inference is the "proliferation of taxonomic categories" (Queiroz and Gauthier 1992, 457). DNA evidence tends to "split" more than "lump" species together into taxon groups. The applications of names to these DNA barcode strings, however, and their reconciliation with existing taxon concepts, is an entirely separate activity, often performed after these phylogenies have been constructed. Increasingly, names are not being applied to this growing cache of genetically labeled information.

Berry van der Hoorn, group leader for biodiversity discovery at Naturalis Biodiversity Center, was kind enough to summarize a case in point (interview 2016). Van der Hoorn's group was in the process of conducting state-funded research on water quality in the Leiden area of the Netherlands. They were charged with identifying species in various water wells in an effort to better understand shifting species variability in different environments. One related project, in particular, took them to an island in the Caribbean, where they were charged with understanding how species variability changed over time. However, because the research in question was about species *change*, the particular species in question, and their Linnaean names, were not a necessary aspect of this work. Van der Hoorn noted, "Sometimes you don't even need to know the species name, you just indicate, we found 150 spiders here, 12 spiders there, and that's enough. Species were identified by 'OTUs'—operational taxonomic units. And you can use [these OTUs] fairly well as a biodiversity index and you don't even need to recognize the species itself." The same was often the case for their study of water samples in Leiden—the actual species involved were less important than the rates of variability and the rate of change for OTUs. For van der Hoorn, the research questions were ecologically based, formulated within and for very specific conditions. The application of names was secondary to solving these project- and funding-specific research queries.

In practice, numerous scholars have pointed to this widening divide between traditional and phylogenetic approaches, and the detrimental effects it is having on the adequate accumulation and collocation of scientific knowledge as each proceeds forward invoking and implementing different methodologies. Nico Franz discusses this increasing tendency to avoid translating phylogenies into classifications in his article "On the Lack of Good Scientific Reasons for the Growing Phylogeny/Classification Gap." Franz notes, "By supplementing a traditional classification with a more precise estimate of phylogeny, one has not yet 'removed the need to use' any or all parts of that classification. In the vast majority of cases, the more recent phylogenetic analyses are properly considered revisions of pre-existing hypotheses (however coarse) about the relationships among taxa and the evolutionary histories of character traits" (Franz 2005, 496). The traditional modes of taxonomy that have been built over the last 250 years, including the application of names to taxon groups, are essential to contextualizing and making meaningful the results of phylogenetic analysis. Additionally, unnamed phylogenies become siloed from this cache of knowledge linked to the historiographical record. One camp cannot communicate with other, thereby limiting the ability for systematics as a whole to proceed forward as a coherent unit and to build on the virtues of each approach.

Despite these competing views, there is certainly a synergy between the two camps that can flourish. A project such as the one described by van der Hoorn provides the raw data that taxonomists can use to produce more robust and complete classifications. Systems like the Barcode of Life are intended to serve this very purpose. In the meantime, however, while these increasing caches of barcodes are being produced, this information fails to make its way into aggregated and composite systems such as the Catalogue. GBIF, for example, is currently working on mechanisms by which genetic barcodes can be appended to their Nub Taxonomy framework. For example, species that are databased can often be recognized at a higher taxonomic level—say, at the order or family level. For example, in Berry van der Hoorn's described project above, spiders were collected from an area with passive traps. GBIF's goal in this case might be to append the barcoded data at the highest known taxon level—in this case, the barcoded spider data might be appended to the Araneae order level of the Nub Taxonomy. The assumption is that data imperfectly placed but available is far preferred over perfectly situated data, especially given that the latter can take a great deal of time to assess. Users seeking out this information then have the ability to search through the GBIF portal to locate this information, even if its exact location on the taxonomic scale is unknown.

As Quentin Wheeler stated, "Phylogenetic classifications are optimal for storing and predicting information, but phylogeny divorced from taxonomy is ephemeral and erodes the accuracy and information content of the language of biology" (Wheeler 2004). A fundamental issue, and one that information studies should be closely attuned to, is how the biodiversity world is attending to these divides and attempting to build classification systems that are more inclusive of multiple approaches to classification based on contradictory standards. The ultimate goal of taxonomies such as the Catalogue are to create inclusive spaces that radically commingle divergent taxonomic opinions. This is no easy feat, and only time will tell in what manner all of these problems will work out within the system. That the Catalogue, and other similar structures, are still vying for broad, universal acceptance is secondary to this story in many ways—that they are trying to bridge divides is what is paramount.

However, even while global data are conceptualized as inclusive of multiple forms in museum and biodiversity practice, much of the networked infrastructure that supports this work is incapable of more complex data sets, particularly video and three-dimensional data (because of both the memory required to store these instances and third-party software requirements to display these assets). For example, the media asset management system in use at the Natural History Museum, London, as of 2016—Open Text Media Manager (known internally as MAM)—does not have the capability to store video or 3D files. As Matthew Woodburn, science data architect for the Natural History Museum's Digital Collections Programme, indicated, "We either have to wait for that capability to be put in place or we need to look for a workaround if it's not going to happen. There are [many] kinds of difficulties on the 3D side of it. Is 3D an image or is it a dataset? Because effectively it's all bits and bytes but it tends to be . . . visualized by a particular piece of software. . . . We can publish it as a dataset, which is not a problem. . . . Then it would be up to them to find a visualization service" (2016). In more than one conversation with museum specialists, a distinction was made between "data that can be used in any particular database" and data "that was unable to be properly ingested into the digitization and digital management program." As such, what constitutes data within an institutional setting at a pragmatic level is often dictated by what formats and types of data any particular system has the ability to accept and integrate.

Lastly, it is one thing to speak about the irreconcilability of Linnaean nomenclature and genetic sequencing data, but it is quite another, more serious, problem to say that the current landscape of biodiversity taxonomy is firmly rooted in a scientific tradition that arose in the context of "the West." This is a fact that has been an undercurrent in some of this book's former chapters—from the exclusion of indigenous information in the classification of the dingo; to biological taxonomy's adoption of Linnaean nomenclature, dependent as it is on Latinate syntactical forms. Further, throughout this text I have spoken about instantiative, instrumental, aesthetic, material, and extensive powers, all of which gain their epistemic strength through the broadscale adoption of colonial reasoning. Speaking with complete honesty, in the many months of fieldwork I undertook to complete this book, the subject of colonial and imperial power as an underlying ethical issue in biodiversity work was the most contentious subject for the many practicing taxonomists I encountered. Many biodiversity workers were keenly aware and conflicted at the imbalance of power their positions represent. Part of this, I believe, has to do with the enormity and pervasiveness of the subject, and the reality that, if one acknowledges it as a problem, then one must also accept the reality that the entire foundation of taxonomy as a field must reckon with the greatest exertion of power of them all: racism and the imperial and colonial motive to dominate. And although this may be true, the reality is that not only taxonomy but also the whole endeavor of science and the Western traditions that matured around it have to come to terms with this reality. Our natural history museums, after all, are full of specimens and material from foreign lands—much of which was secured during the great colonial expansion of the nineteenth century. If there are "blind spots" for data, chances are they are in areas that have traditionally been exploited and underrepresented on the global stage. I am thinking here of the gap in African data that the Catalogue of Life has been actively trying to rectify. In chapter 8, I take a sharp critical turn and delve more deeply into this issue and frame the work of classification as unescapably intertwined with these historical realities.

8 POWER OF COLONIALITY AND A MOVE TOWARD JUSTICE

THE DARKER SIDE OF CLASSIFICATION: FROM *A'NINANDAK'* TO BALSAM FIR

I have tried to make the case that derivative positionality (within classifications) is an important mode of power in classifications. How one's identity (human and nonhuman alike) is positioned in a classification has material and aesthetic effects that radiate outward into social spaces. Position *is* power, I have tried to argue, and have tried to show how power emanates from the opaque and systemic qualities inherent in classifications. The more accurately an identity (of a person, a thing, an organism, and the like) is expressed or reflected in systems, the more we can see that system as being *just* in relation to a lived reality. We concluded chapter 7 with a section titled, "Divergent Traditions and Nameless Taxa," which emphasized the structural problems associated with phylogenetically derived classifications. Such genetic classifications pose structural, syntactic, and semantic problems, as they are not always matched with prevailing Linnaean nomenclature. This means that phylogenetic information is effectively precluded from systems that use binomial nomenclature as the primary collocating mechanism—which at present is most, if not all, of the classifications that dominate the discipline. The result is that two parallel methods of classification are being practiced, each producing valuable biological and ecological information, but their paths cannot cross or intermingle in any pragmatic way. Surely this is problematic. However, if we wished, we could hypothetically imagine hiring hundreds of taxonomists to apply

Linnaean names to genetically derived operational taxonomic units. The same cannot be said for some forms of knowledge that are ontologically or epistemically on different planes from the structures of Western science. So here we turn to the fundamental problem with *all* classifications within the colonial episteme.

The question of structural power goes well beyond the internal components of classification—which is to say, a classification's technical specifications, policies, and practices—it extends outward to the epistemic structure of *how we collectively know what we know at all.* And herein lies the true clutch of power: for all our discussion of composite systems, their epistemic borders are defined by particular ways of knowing—Western ways of knowing. And these boundaries are nonnegotiable. This shouldn't come as a surprise to anybody. By and large, the impulse, trajectory, and method of scientific discovery are distinctly Western practices. Historically, the genesis of natural history as a discipline—and later the biological sciences—was the imperial drive to collect "foreign" species from colonized lands, which were then collected, studied, and displayed within personal collections as cabinets of curiosities. Later these often-private collections ended up in museums, where the "control" of nature can be on full display (Findlen 1994, 153; Raby 2017, 207). Even the deeply influential thinking of Charles Darwin precipitated from British expeditions intended to survey the land, sea, and air in various permutations, in an attempt to create a more unified and global view of the natural world (Richards 1993, chap. 2).

I do not want to imply that the function of the Catalogue of Life was in any way meant to somehow solve the problem of coloniality—it simply cannot. All the Catalogue ever promised was to provide a mechanism for data control and transfer to facilitate access to the broadest extent of biodiversity knowledge available. But what is intriguing about a management classification is that it forces us to question the essential functions and problematics of standardized, normalized spaces. As written by Bengt Jacobsson, "Standards are created by groups of people who develop solutions which they regard as good for all concerned. . . . A significant feature of standards and standardization is that expert knowledge is stored in *rules* and in technical solutions. Knowledge is transformed into rules that are abstract, general, and recorded in writing" (2005, 41). The more we universalize and aggregate, the faster we lose sight of local nuance and the further away "from nature," or from the local site of empirical activity, our research and knowledge becomes. One cannot get more "local" than the indigenous groups forcibly relocated throughout the world. The conflict in the biodiversity taxonomic community with respect to management taxonomies is instructive for all of us that struggle daily to meet the needs of diverse constituents. It brings to light the core material and epistemic realities of our organizational work. The perennial question remains: Do we allow the knowledge of our world to remain fragmented to ensure local integrity, or force these systems together in the interest of global progress and international cooperation? Perhaps most will say that we can have both. If we do, then the Catalogue is instructive precisely because it displays the benefits and pitfalls of such an approach. We cannot, it seems, have our cake and eat it, too, however, even coupled with the technological capacities we have at our disposal. Neither approach is perfect—global aggregation or fully local systems—especially while there are largescale issues such as global climate change that require the aggregation of data. What I *can* say is that the solutions put forth by the Catalogue in terms of knowledge dissemination only make the need to push for more inclusive, anticolonial projects all the more pertinent. The cost for moving away from the local is that large-scale infrastructures only magnify the cultural harms that have historically been distributed, making it much more difficult to intervene and ameliorate these harms. Such is the problem with systemic power.

So, while in our contemporary world we see the endeavor of biodiversity science as a stream of inquiry driven by the pure pursuit of knowledge, it is impossible to divide these ambitions from the colonial impulse to collect the extraordinary—which is to say, to own the knowledge of foreign lands for the pursuit of local invention and exploitation. A discussion of structural and epistemic power cannot be complete unless we formally acknowledge these pervasive, historical qualities of classifications. And indeed, this colonial impulse remains even into the nineteenth, twentieth centuries, and twenty-first centuries, as evidenced by Megan Raby in her *American Tropics: The Caribbean Roots of Biodiversity Science* (2017). As conveyed by Raby, "U.S. biologists became embedded in the networks of empire. . . . This came in many forms: interimperial solidarity with the British and West Indies, ties to the U.S. government power structures in Puerto Rico or the Panama Canal Zone, or reliance on land-owning U.S. corporations and private individuals" (2017, 216). A contemporary notion of biodiversity, then, cannot be severed from historical origins, ingrained as they are in regimes of power and global control.

It has been a long time coming, but the discipline of information studies and, more specifically, the subareas of classification and representation have been fortunate in having a host of scholars specializing in indigenous and post / decolonial thinking (Bone and Lougheed 2018; Corn and Patrick 2019; Duarte and Belarde-Lewis 2015; Littletree, Belarde-Lewis, and Duarte 2020; Montenegro 2019). As characterized by Duarte and Belarde-Lewis,

> Broadly, colonization—the verb, or enactment, of colonialism—is based on four overlapping mechanisms: (1) the classification of diverse Indigenous peoples as a single lesser-class of sub-humans deserving of social subjugation at best and extermination at worst; (2) the theft and settlement of Indigenous lands and social spaces by an elite Settler class; (3) the articulation of institutions to support this class system and the elite control of the environment; and (4) the disciplining of elite forms of knowledge through the marginalization of Indigenous. (2015, 682)

Expanding on the fourth quality of colonization, then, let us examine the hold that coloniality has on the process of classification and representation in the biodiversity taxonomy. In light of this, a central goal in deconstructing the power of classification systems is to imagine ways by which we can break free from the colonial constructs that preclude the inclusion of knowledge that falls outside its epistemic boundaries. This problem must go beyond merely pluralizing the classificatory record; it forces us to imagine what it means to acknowledge these colonial harms and envision ways of designing systems that ameliorate these harms. And to a certain extent, perhaps this may not be wholly possible—there will never be one system that can attend to all cultures, and I seriously doubt we can design a system in the West (or anywhere else, for that matter) that is completely unburdened from colonial modes of knowledge. In these cases, then, we have to design systems that actively distribute power back to the cultures that were exploited for their indigenous and endemic knowledge.

In information studies, Hope Olson is recognized for reminding us that subject representation in library systems matters, and that language is both a necessary vehicle for access and a homogenizing agent that obstructs subject complexity in smoothly functioning systems that construct seamless classificatory realities (Olson 2002, 238). But even in Olson's sense, however we might express the relationship of a "subject" to its represented "lived complexity" in language, these expressions are still expressed from within a fairly limited epistemic framework. Our goal with classification, in general, should not be to merely expand the universe of subjects to widen a classification's expressive capacities (though that is certainly a good start!); we should also push to define our methods and theories in new terms—away from the univocal narrative that these classifications have represented for far too long.

The co-opting of knowledge from indigenous tribes, for example, has historically been gift-wrapped in discourse that presents these "communicative" actions as a worthy and ethically sound expansion of our collective knowledge. The assumption is that we are the better, scientifically and socially, for being inclusive to alternative forms of knowledge. However, what we often fail to realize is that, historically, the modes in which we collect this knowledge, and the information structures that we embed it within, erase any supposed "progress" we might make to expand and pluralize the boundaries of our collective intellectual spaces.

A case in point is the US effort to collect indigenous knowledge through ethnographic means in the late nineteenth and early twentieth centuries. A key moment in this push to collect indigenous knowledge was the establishment of the United States Bureau of Ethnology (BE) in 1879. The bureau's initial goal was to manage the archival materials produced by the US Department of the Interior, which included an increasing number of ethnographic surveys of indigenous tribes dating from 1867 to 1874

(Woodbury and Woodbury 1999). By 1894, however, in acknowledgment of the geographic scope of the bureau, its name was changed to Bureau of American Ethnology (BAE); under the direction of John Wesley Powell, it began to more concertedly embark on organizationally funded ethnographic work, especially emphasizing the documentation of the American tribal languages. Over time, the scope of the project expanded to include a fairly broad range of collected knowledge, including religious and ceremonial practices, mythology, burial practices, art and metallurgy, textiles and pottery, music, and architecture. The data from these long-term expeditions and ethnographic studies were published in the *Annual Reports and Bulletins of the Bureau of American Ethnology*, which, although relatively little-discussed today in IS, still remain a crucial historical repository that expresses the powerful colonial impulse for the appropriation of knowledge. For our purposes, the ethnobotanical and ethnozoological surveys are of particular significance, because they give us a very clear picture of how detrimental the "translation" of native knowledge into Western forms truly was.

In an article titled "Use of Plants by the Chippewa Indians" (Smithsonian Institution Bureau of American Ethnology 1895, 286), written by ethnomusicologist Frances Densmore, we can see how the bureau collected the native names of plants, documented their usage, and then mapped those indigenous names to the common and official scientific binomial then in use by the scientific community (see figure 8.1). In figure 8.1, we can see Densmore's documentation of *a'ninandak'*, which has been mapped to the balsam fir, *Abies balsamea* (L.) Mill (which, incidentally, is still the accepted scientific name for the balsam fir). Alongside these entries, we see specific notes about *a'ninandak'*, regarding how the Chippewa used this plant for medicine, food, and utility, as well as cross-references within the *Bulletins* that show how other tribes in the area also used these items. Pages and pages of entries follow, listing the specific plants used for medicinal purposes, as well as the active chemical agents within these specimens that contributed to these pharmacological effects. A huge boon to US pharmacology to be sure, which Boaventura de Sousa Santos and Vandana Shiva (2008) make clear in their discussion of the colonial act of patenting of biodiversity as a mode of biopiracy.

286 USES OF PLANTS BY THE CHIPPEWA INDIANS [ETH. ANN. 44

LIST OF PLANTS ARRANGED ACCORDING TO BOTANICAL NAME

Botanical name	Common name	Native name	Meaning	Use	Reference to use by other tribes[1]
Abies balsamea (L.) Mill.	balsam fir	a'ninandak'		medicine (headache)	
Acer saccharum Marsh.	sugar maple	a'nina'tig		food, utility	
Achillea millefolium L.	yarrow	a'djidamo'wano	squirrel tail	medicine (headache)	swelling, etc., Winnebago, 33d Rept. B. A. E., p. 134.
Acorus calamus L.	calamus	wikĕn'			
		na'bugûck'	something flat	medicine (cold, etc.)	fever, cough, etc., 33d Rept. B. A. E., p. 69.
		mûckosija'bosigûn	hay purgative	charm	
Actaea rubra (Ait.) Willd.	red baneberry	wi'cosidji'bik	drawing root or plant	medicine (diseases of women).	
Agastache anethiodora (Nutt.) Britton.	giant hyssop	weza'wûnûckwûk'	yellow plant	medicine (cough and pain in chest).	food, 33d Rept. B. A. E., p. 113.
Allionia nyctaginea Michx	umbrella-plant	be'dukadak'igisin	"it sticks up"	medicine (sprain)	fever, etc., 33d Rept. B. A. E., p. 78; fracture, Sioux, Bull. 61, p. 261.
Allium stellatum Ker.	wild onion	mûckode'cigaga'wûnj	prairie skunk plant	medicine (colds)	
Allium tricoccum Ait.	wild leek	siga'gawûnj'	onion	medicine (emetic)	
Alnus incana (L.) Moench.	alder	wadûb'		medicine (diseases of women), dye.	
Amelanchier canadensis (L.) Medic.	shadbush	guzigwa'kominaga'wûnj	thorny wood	medicine (dysentery, diseases of women), food.	
Anaphalis margaritacea (L.) B. & H.	pearly everlasting	wa'bigwûn	flowers	medicine (paralysis)	
Andropogon furcatus Muhl.	bluestem	mûckode'kanês	small prairie	medicine (indigestion)	fever, etc., Omaha-Ponca, 33d Rept. B. A. E., p. 68.
Apocynum sp.	dogbane	beba'mokodjibika'gisin	"bear root, it is found here and there."	medicine (cough)	
Apocynum androsaemifolium L.	do	sasa'bikwan		medicine (heart palpitation, earache, headache; a baby's cold; also for charm).	
		ma'kwona'gic odji'bik	bear entrails root.		
Aralia nudicaulis L.	wild sarsaparilla	wabos'odji'bik	rabbit root	medicine (remedy for the blood, also applied to a sore), charm.	

Figure 8.1
"Use of Plants Arranged According to Botanical Name," by Frances Densmore. Annual Report of the Bureau of American Ethnology to the Secretary of the Smithsonian Institution 44th (Smithsonian Institution Bureau of American Ethnology 1895).

The organizational context for the collection of this knowledge was, to put it mildly, tinged with the darker intents of these expeditions. Top US anthropologists were recruited by the BAE under the premise that the "settling of the West would bring an inevitable end to the primitive lifeways [of American tribes] that had remained unchanged throughout the centuries" (Woodbury and Woodbury 1999, 285). And even worse, when the bureau was founded, John Powell's initial task was to classify the US Indian tribes according to levels of "cultural sophistication" (Woodbury and Woodbury 1999, 284). Powell, subscribing to the theories of Lewis Henry Morgan, categorized all US native tribes as "savage" in the taxonomy of human advancement, with the exception of the southwestern Pueblo peoples, who were graced with the classification of barbarism, one notch above savagery. That this was the common characterization of US native peoples is secondary to the fact that there is no historical contingency imaginable that merits this kind of cultural subjugation. The collection of knowledge was less about the epistemic expansion of knowledge and more about the collection of knowledge-as-resource that was quickly dissipating because of the damaging US policies that were to eventually (nearly) eradicate the widespread existence of native knowledge. The knowledge, translated, was integrated into, and became property of, the US episteme as conveyed through governmental organizations.

And indeed, as we might guess, this explicit imbalance of knowledge power is expressed in the bureau's classification of indigenous knowledge itself, which falls into the trap of Western scientific knowledge forms. As Wendy Geniusz (2009, 6–7) notes, "Although the Anishinaabeg [in English, the Chippewa, Ojibway, Ojibwa, or Ojibwe] have stories, religions, music, and botanical information, these are not the extremely specialized narrowly defined categories of non-native scholarly work. Within the Anishinaabeg cultural context, one does not ignore information in one of these categories in order to concentrate exclusively on another." The reductionist model that much of scientific activity depends on is counter to the holistic, intersectional, and temporally contingent constitution of knowledge within indigenous cultures. Such a fact is true in many contexts, even beyond the Chippewa, for we see the importance of story in indigenous cultures throughout the world. So, when Hope Olson declares that naming *matters*, the kind of names she refers to (that of the subject within catalogues) are of no great consequence in the world of indigenous tribes. Joeliee Seed-Pihama (in Archibald, Lee-Morgan, and De Santolo 2019) describes how names in Māori cultures are important, yes, but perhaps one's name is less important than the story of where it is from. And when a name is important, context is everything. "Our genealogical connection with all phenomena in the universe intricately interlinks us," Seed-Pihama states, "much like a spider web. Therefore, Māori never act and see things on an individual level" (Archibald, Lee-Morgan, and De Santolo 2019, 107).

Of course, names in biodiversity also have a context that is important, but that context is also derived through traditional scientific practices. To manufacture a taxon concept, as we saw, is to invoke a whole host of sources that bring that concept to life. These concepts are ultimately tied to nomenclature, which makes them tractable in discourse and databases. But Seed-Pihama's sense of context is far more contingent, historically attuned, and intermingled with use, place, and cosmological identity. We see inklings of this understanding in the *Bulletins* as well. Frances Densmore continues her narrative about the Chippewa by saying,

> There is no exact terminology of Chippewa plants, although there are some generally accepted designations of common plants and trees. In obtaining the names of plants it was found that the same name is often given to several plants, and that one plant may have several names. Individuals often had their own names for the plants which they used as remedies. It was also customary for a medicine man, when teaching the use of a plant, to show a specimen of the plant without giving it any name. (Smithsonian Institution Bureau of American Ethnology 1895, 297)

We can see how the way we understand *relationships* between terms in classifications is fairly limited when juxtaposed with the complex associations of names and organism designations within the Chippewa culture. How does one convey a story and its multifold semantic relationships in classification space, for example? And to be fair to any one classifier, at the present we can't technically program our systems to accept these extra-epistemic associations, especially if our culture fundamentally (historically and on contemporary terms) fails to see them as significant to the understanding of the natural world. In a statement that will forever resonate with me, Seed-Pihama unknowingly echoes Olson when she states, "The renaming of our people was intentional and purposeful on the part of the colonizers. The *power to name*, or rename, is a specific kind of symbolic violence and power that superimposes and defines what is seen as accepted as normal and legitimate within society" (Archibald, Lee-Morgan, and De Santolo 2019, 117; emphasis added). The "power to rename"—and to rename over and over and over—should be the new mantra of IS scholars examining colonial and imperial appropriation of knowledge.

Acts such as this appropriation of knowledge (renaming in its own right) is what Walter Mignolo was critiquing when he published *The Darker Side of the Renaissance* (2003) and *The Darker Side of Western Modernity* (2011). In both of these texts, but especially *The Darker Side of the Renaissance*, Mignolo points to the impact of colonization on our modes of knowledge production. Part and parcel with the process of colonization was the representation and documentation of the indigenous "other" into forms that conformed to Western standards—a standardized form that expresses what Mignolo later calls the colonial matrix of power. This is not altogether surprising given that, according to Mignolo, "the organization, evaluation, and transmission of a *set* of events as *historical* events are in large scale dependent upon the rhetorical restrictions of narrative genres as well as on the skills of the person narrating them, in oral or graphic form" (Mignolo 2003, 178; emphasis original). If indigenous ways of knowing do not align with the epistemic orientations of the classifier, then it is impossible for us to expect that the outcome of the process of classification would be anything other than suppressive to more nonlinear and associative ways of knowing. Indigenous knowledge in this sense is conditional. As Johanna Drucker writes, "A conditional document is not a speculative one, not imaginative or imagined, but is produced by protocols and processes that use structured conditions as a way to run, operate, select information, and display it" (2014a, 25). If narrative, historicity, and atemporality partially define an indigenous way of understanding and identifying nature, then we cannot expect to do knowledge justice in structurally absolute and categoric hierarchies. The shadows of colonialism will always be present in our forms of knowledge, unless we redefine our practices with radical new ways of expressing connectivity.

As Linda Tuhiwai Smith notes in her influential *Decolonizing Methodologies* (2012), the Western notion of research implicitly distances the "human" from the process of empirical examination. To examine nature, we measure it, weigh it, and rely on other empirical variables to understand how the object of interest somehow relates to intellectual aspects of our social world. The human stands apart from nature, but is always, in the end, at the center of its annunciation and distribution. Deductively disembodying the components of nature merely to reassemble them in classificatory structures is an act that's foreign, and in some sense violent, toward the more holistic and harmonious understandings of nature that are present in indigenous societies. In Marisa Duarte's *Network Sovereignty: Building the Internet across Indian Country* (2017), the author shows the complex and reciprocal relationship between information communication technologies and indigenous cultures and how the imposition of these technologies requires the rearticulation of indigenous values as well as a decentering of Western practices toward the decolonization of indigenous lands, people, and knowledge. To partly frame this analysis, Duarte invokes a coloniality of power analytic that shows how classifications, institutions, space, and knowledge connect in a circular pattern of authority that slowly degrades the epistemic authority of indigenous cultures (2017, 18–25). The framework illustrates how power is systemically derived and distributed, and also how that power becomes part of indigenous thinking, thereby maintaining imperial and colonial power by reformulating the epistemic commitments of indigenous communities with the values and practices of Western culture. The story of the dingo conveyed earlier in this book describes exactly why: even while the government allows toxic baiting, indigenous populations wish to save the dingo as sacred, integral to their cultural storytelling. Yet in order to offset and push against these realities, indigenous communities must enter the discourse of Western politics to remedy any ill effects of dingo legislation. These are the stories we need to attend to more as we push for widescale biological species conservation in light of global tragedies such as climate change. In practice, we have to engage with decoupling method from design—is it is possible to design with new methods that push against prevailing normative and destructive colonial frames? To narrow the gap between the design of systems and their usually violent implementations, such that local communities have a say in how they are expressed.

As classification designers, we need to prioritize better understanding these darker colonial, Western, and imperial realities reflected in the narratives of classification and representation. But we cannot only acknowledge them, we must devise new ways of representing knowing, order, and disorder. We can never attain the true extent of epistemic justice until we reckon with these historical realities. If we cannot set aside our imperial scientific foundations (and I am not sure that this is even a feasible action) or integrate information in more culturally sensitive and humane ways (and I certainly think we can and must), then we need to fundamentally reformulate how we envision our current forms of taxonomic knowledge. We need to locate means of representing more complex, conditional, and networked ways of knowing, rather than ones of hierarchy and determinacy that necessarily express power imbalances. Our narratives should re-center processuality and indeterminacy as a narrative technique, particularly because it can help users of biodiversity information understand that taxonomies, while informed, are not static and free of divergent opinions. And even if we can't eschew ourselves of taxonomic hierarchies, given the limitations of our technologies or graphical spaces in general, then we need to introduce mechanisms that reveal transparency where there is opaqueness, and contingency where there is rigidity.

For these reasons, more often than not, I side with Maurizio Ferraris and Richard Davies when they embrace and advocate for richer ontologies and the exemplification of *more* objects over a more stripped-down parsimonious view of the world (2013, 15–16). If we do this in the domain of classification building, what we initially lose in precision we can gain in diversity. Is not the documentation and emergence of new, more comprehensive forms of knowledge an avenue toward a more precise informational world, in any case? Let us redefine precision. If we embrace the reality that, when we *design* systems, we are also constructing ways of being in the world, then perhaps the stakes of our intellectual work will become more apparent, and these stakes will begin to influence our practice (Escobar 2018, 4).

PLURIVERSALITY AND DESIGN

Attempts to bring pluriversality into practice have been slow in coming—partly because such approaches sidestep the problem of facilitating what we've been calling global communication—and certainly some of the aims of this communication are worth pursuing! Climate change is real, and the solutions for such a phenomenon are global and not local. But nonetheless, such a reality does not mean that, in tandem, we cannot pursue a plural, classificatory ideal. At the center of this book has been a composite method of classification. This mode of organization is spurred, at least primarily, by the increasing rate of biodiversity data being produced around the world. However, for all we have gained in terms of a global view, these structures also remove taxonomic knowledge from the local site of production and have not yet fully realized how to maintain that global-local connection. While we want our data to represent the global, we have to push against the simultaneous reality that it is also being globalized, in the capitalist sense that global gains come at the cost of local communities and their well-being.

Within this context, I am defining pluriversality as the embrace of many classification authorities, and positing that propagating the many is more valuable than manifesting the universal singular. Speaking within the context of political universality, Judith Butler notes how, "the use of the doctrine of universality" has been invoked "in the service of colonialism and imperialism" (2000, 14–15). As has been indicated on numerous occasions in this book, universality (as evidenced in classifications, both traditional and composite), whether purposeful or not, acts as a passive tool of knowledge control and, by extension, a mechanism of intellectual and social power. "The fear of course," remarks Butler, "is that what is named as universal is the parochial property of the dominant culture, and the 'universalizability' is dissociable from imperial expansion" (2000, 15).

The ramifications of this kind of embedded colonial and normative power, while certainly detrimental and violent within the social sphere, have proven particular devastating to the natural world. The 1992 United Nations Rio Declaration also identifies the importance of indigenous knowledge and the need to reverse these colonial erasures. Article 22 states,

> Indigenous people and their communities and other local communities have a vital role in environmental management and development because of their knowledge and traditional practices. States should recognize and duly support their identity, culture and interests and enable their effective participation in the achievement of sustainable development. (Convention on Biological Diversity 2006, article 22).

We cannot continue to see ourselves as apart from the world of nature, climate, and environment, and continue to couch these phenomenon as if they are "happening to us." They fundamentally *are us*, not only because we caused them (though that is also true), but because we, literally, cannot live without them.

In *The Disorder of Things*, John Dupré advocates for what he calls a promiscuous realist approach to classification. To subscribe to promiscuous realism is to hold a "radically ontological" pluralist claim that there are "many equally legitimate ways of dividing the world into kinds" (Dupré 1993, 18, 11). Dupré notes, "I can see no reason why commitment to many overlapping kinds of things should threaten the reality of any of them. A certain entity might be a real whale, a real mammal, a real top predator in the food chain, and even a real fish" (1996, 262). The danger of such an approach is obvious: we, of course, know that not everything is legitimate—after all, the International Flat Earth Research Society is a thing in this world some people see as legitimate. And in some cases, the beliefs of the few can come at the cost of the many: the increasing number of anti-vaccination supporters promises to continually extend what has been a horrendous and deadly COVID-19 global pandemic. To embrace classificatory plurality is not to eschew the concept of rightness and justness. To be pluriversal is to embrace a process that strives to better understand divergent modes of *knowledge production*, not necessarily all said knowledge that these accepted processes produce—a nuanced but important difference. Every culture has their epistemic standards. Mignolo writes,

> Pluriversality (or diversality, in the sense of diversity and not in the sense of dichotomy) is not the rejection of universal claims. It rejects universality understood as abstract universal grounded in mono-logic. A universal principle grounded in the idea of the diversal (or pluriversal) is not a contradiction in terms, but rather a displacement of conceptual structures. Diversity as universal is . . . a world composed of multiple worlds, *the right to be different* because we are all equals. (Mignolo 2011, 234–235; emphasis original)

In this book I have discussed process theory, for example, and how the process of dividing nature into entities such as species taxon concepts is tantamount to imposing order and boundaries where it otherwise does not exist. To embrace pluriversality in this broader way is to acknowledge, on the one hand, that we *must* manufacture entities in this way—this is how we think, the way we construct language and function semiotically, and also how we will combat large scale catastrophes that science must continue to inform—but, on the other hand, we also have a responsibility to balance this process of compartmentalization by adopting and implementing technical mechanisms that convey the limitations of said knowledge within certain use contexts. Again, the Catalogue, on the face of it, is worthwhile, but the problem remains that its infrastructure is not fully able to directly and explicitly convey its limitations to nonexpert users.

One way this transparency can be implemented is by practically activating what Arturo Escobar calls a "radical relationality" (Escobar 2020, xiii). Classifications should be cohesive but also attend to local conditions as needed—and the mechanisms for such transformation should be visible and alterable. A radical relational approach is one that posits that all entities (human and nonhuman) are "so deeply interrelated that they have no intrinsic, separate existence by themselves." What we separate, we can again link together into a cohesive unit. So a deeper question becomes, How can our taxonomies present themselves as a unified whole, while also expressing the multiple other arrangement that ostensibly could be? This is important, not for the biodiversity scientists who understand the hypothesizing nature of their craft, but for the general public, who are not qualified or acquainted with biological taxonomy enough to question the arrangement of any one structure. As previously mentioned, ontologies are certainly one way we can come close to modeling multidimensional relationships, though even these models are limited, based on intended use and function and not without their own drawbacks, as Nico Franz has outlined (2010).

Jens-Erik Mai embraces a similar approach in his "The Modernity of Classification," when he says, "Plurality is not something that can be set aside as simply something that has do with culture, society and language, but it is also something that has to do with the individual. What is most important, and, perhaps its most enduring quality is of locality: justice to the particular, the specific, the located" (2011, 723). To embrace plurality is to push against the ongoing debates about what constitutes right and wrong mechanisms for the organization of entities. This seems to me a much more practical way of understanding the potentiality of classificatory spaces. If we see classifications as the contingent mechanisms they are, they have far less power over the definition of our identities and our (and other entities) situated place in the world.

The difficulty is, and always will be, imagining how pluriversality is to be translated into the *design* of systems. This problem of implementation comes with no obvious solutions. One way forward, perhaps, is to imagine a system that defines the success of usability not as consistency, speed, and accuracy (whatever these may mean in practice within a particular context), but rather as flexibility, care, and the active display of responses that emphasize context and difference. Information philosopher Kay Mathiesen's (2016) work provides a nice jumping-off point for such an approach. To attain informational justice, Mathiesen argues, we must account for three basic forms of i-justice: iDistributive justice, iParticipatory justice and iRecognitional justice. The premise is that information systems should allow for the equitable and fair distribution of information; that the environment for system design should allow for collaborative and shared decision making about what is best for any individual or group; and that the information in said system should justly represent members of society. In light of this, we can see another reason that large-scale infrastructure often falls short of its epistemic responsibilities: it is too far removed from the constituents of interest. Once again, the local provides avenues for more robust collaborative and community integrative mechanisms.

The first step, it seems to me, is to acknowledge that, if our representation systems have material effects on the social world, then the solutions that we apply to our social spaces should also be reflected in the systems we create. As we grapple with rising temperatures and declining biodiversity, we know that unless we change our daily practices (and corporations are forced to implement large-scale change), the future of our society is doomed to fail as our ecological spaces crumble. The colonial, industrial, and capitalist epistemes that we have long subscribed to are now being seen for the detriment that they are. And, as such, we should equally see that the intellectual frameworks, tools, and repositories of knowledge that have supported these initiatives must be dismantled and rebuilt. If we have to change our practices in the environment to meet the challenges these crises present, then we must also change the space of systems that make it so. This is why Lefebvre's triad is so important. *All* space is connected—from the material, to the represented, to the imaginary, and back again.

I see some promise in transitional discourse approaches to design, which principally note the deep connection between the environment and social spaces, and take this reality as a jumping-off point to resituate how we must radically transition our practices to ones that are environmentally sustainable—especially if we understand this model as the practical implementation of Mathiesen's model of informational justice and the design ethic that build's off Ostrom's SES examination of multicommunity systems that was introduced in chapter 3. "Transitional discourses," Escobar notes, "take as their point of departure the notion that contemporary ecological and social crises are inseparable from the model of social life that have become dominant over the past few centuries, whether categorized as industrialism, capitalism, modernity, (neo)liberalism, anthropocentrism, rationalism, patriarchy, secularism, or Judeo-Christian civilization" (Escobar 2018, 139). To acknowledge that we, as well as the environment and the systems that support it, are in transition is to expect that no radical change to the system is going to happen overnight. Rather, change is emergent and iterative and reveals itself in the process of redesigning our systemic potentials.

In *Designs for the Pluriverse: Radical Interdependence, Autonomy, and the Making of Worlds* (2018, 152–164), Escobar outlines the transition design framework at Carnegie Mellon University's School of Design. The design program, focused as it is on producing "system thinkers" that integrate exceptional change in real-world design applications, focuses on design of and within new paradigms. Particularly useful, I think, is the four-stage Transition Design Framework that lays out the process involved in such an approach (see figure 8.2). Stage 1, Visions of Transition, requires a clear articulation of a particular need at hand within society, including, among other areas, a formulation of "information culture" such that it is more in harmony with our surrounding ecosystem; Stage 2, Theories of Change, supports the integration of many fields to better understand the dynamics involved with the particular transition at hand; Stage 3, Posture and Mind-Set, pushes for the individual reconceptualization of the possibilities within design, ones that involve the broadscale integration of multiple voices and collaboration; and Stage 4, New Ways of Designing, involves implementation of new systems and the realization that such implementation is iterative and dynamic. What I find most appealing about this approach is that, although transition is disruptive, it need not occur at a rate that destroys the credibility of a system. And further, the design framework works at a scale that would work exceedingly well at the local level of implementation.

What the framework facilitates is a multidisciplinary approach to classification building that embraces the authority of many varieties of expertise. For example, if we look to the design of a biological taxonomy, we can envision the collaboration among taxonomists, ecologists, ontologists, information scientists, and digital humanities and representational experts, as well as the community members who represent diverse epistemic orientations. This kind of experimentation is essential. Importantly, this is not a mechanism by which we can circumvent the expertise of any given taxonomists—quite the contrary, in fact. The goal here is rather to design systems that formally display the nuance of their hypothesizing work, both on an individual level and in relation to the entire field of taxonomic

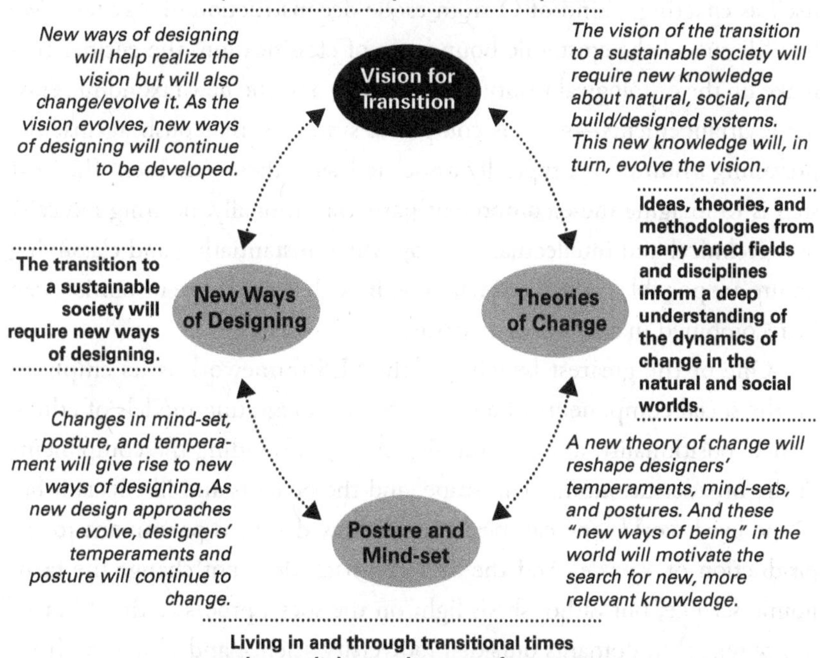

A vision for the transition to a sustainable society is needed. It calls for the reconception of entire lifestyles that are human-scale and place-based but globally connected in their exchange of technology, information, and culture. It calls for communities to be in a symbiotic relationship with their ecosystem.
New ways of designing will help realize the vision but will also change/evolve it. As the vision evolves, new ways of designing will continue to be developed.
Vision for Transition
The vision of the transition to a sustainable society will require new knowledge about natural, social, and build/designed systems. This new knowledge will, in turn, evolve the vision.
Ideas, theories, and methodologies from many varied fields and disciplines inform a deep understanding of the dynamics of change in the natural and social worlds.
The transition to a sustainable society will require new ways of designing.
New Ways of Designing
Theories of Change
Changes in mind-set, posture, and temperament will give rise to new ways of designing. As new design approaches evolve, designers' temperaments and posture will continue to change.
A new theory of change will reshape designers' temperaments, mind-sets, and postures. And these "new ways of being" in the world will motivate the search for new, more relevant knowledge.
Posture and Mind-set
Living in and through transitional times requires a mind-set and posture of openness, mindfulness, a willingness to collaborate, and "optimistic grumpiness."

Figure 8.2

"CMU's Transitional Design Framework" (Escobar 2018, 155).

Sources: Terry Irwin, used by permission; Arturo Escobar, *Designs for the Pluriverse*. Copyright 2018, Duke University Press. All rights reserved. Republished by permission of the copyright holder. www.dukeupress.edu.

practice. The result of transition design is that it embodies "not just the communion of people, their artifacts, and nature, and will come into being at multiple levels of scale" (Escobar 2018, 155–156). Working within this framework, taxonomies can be produced within the context of an intersectional world with many competing levels of interest.

But while the transition design framework is strong theoretically, what it lacks is a clear mechanism by which we can deconstruct a system such that we can reimagine its parts in new, combinatorial ways. If we look back to our discussion of the social-ecological framework (SES) as advocated by Elinor Ostrom (2009), we can begin to see how such a transitional reimagining might be possible. In some ways we might see this book itself as enacting a kind of Ostrom-esque deconstruction of a system. We have discussed the epistemic boundaries of classification, the entities that make up the ontological reality of the system, and the new taxonomic epistemic arrangements—such as composite systems—that push against the prevailing assumptions typically associated with these activities. The next step is to imagine these component parts transitionally, positing radically new technical and intellectual ways by which instantiating and classifying nature are possible, as well as imagining how the entities we instantiate can be recombined in anti-normative ways.

One of the greatest benefits of the SES framework is its emphasis on the social components of a given system, integrating models of ethics, justice, positionality, and community. By understanding the components of the broader taxonomic landscape, and the benefits and ill effects it has in the lived world, one can begin to see how design is paramount to the production of science. And the SES approach does not change the taxonomic science, but rather sheds light on the social processes that limit its full potential in domains outside biodiversity science and systemics. If we understand the social components of taxonomic work, then we can gain an understanding of who should be involved in the iterative transition design. The next steps toward the future of classification practice are ones geared toward justice and environmental benefit.

CLASSIFICATION AS ENVIRONMENTAL AND ECOLOGICAL JUSTICE WORK

Part of a new, transitional mode of thinking is accepting that every*body* (person and organism, alike) has the foundational right to control how their identities are represented in a system. We must acknowledge that classifications are not inert tools, but active epistemic formulations that carry cognitive and affective weight in the world. I want to end this chapter with the notion that, going forward, just as classification work has been framed within the IS domain as an expression of social justice work, the work of biodiversity taxonomists shows how such knowledge systems are also directly engaging with environmental and ecological justice work. If we are to emphasize justice as a crucial aim of the information professions (and I think that we all should orient ourselves toward this directive), it is incumbent on IS professionals to imagine new classificatory frontiers to better understand our impact in all disciplinary spaces. This is to say once again that, for classifications to be *just* entities in the world—fair to the individuals and groups that they represent, as well as morally restorative—they must attend and adequately respond to Miranda Fricker's (2009) injunction against epistemic injustice. Classification and representation systems should support one's pursuit to better conceptualize their position and experience in society and, by extension, should also provide true information that then increases one's capacity and credibility as a speaker. Similarly, that information should support attempts to radically intervene into our ecological processes, such that we reduce our impact on the environment, as well as eradicate the injustices that all living organisms are experiencing. Just as one has a basic right to live in a just world, all living entities should also have a right to access descriptive systems that represent that ideal justness. In short, information is formative, and foundational justice works on all levels.

This critical approach to classification is exactly what Jonathan Furner called for at the 2019 International Society for Knowledge Organization Annual Meeting in Porto, Portugal (Furner 2018). In, "Truth, Relevance, and Justice: Towards a Veritistic Turn for KO," Furner advocates for a new priority in the knowledge organization domain—one that prioritizes truth over relevance as a core goal of classification builders. Furner's new critical approach to classification and knowledge organization, then, focuses on three other goals, including embracing KO systems that are (1) informed by applied social epistemology, such that we interrogate the qualities and arguments that we, as KO specialists, use to justify what is true or false; (2) inspired with values of epistemic justice, in that we foster the dissemination of true beliefs; and (3) respectful of human rights, in that we recognize the need to respect and lend credibility to underrepresented groups that have heretofore been overlooked by systems (which I take, in this context, to include the natural world as well).

In tandem with this, it is important to acknowledge that there are two sides to justice that must be addressed within these systems, at least as they apply to the domain of biodiversity studies as it intersects with classification production. On one hand, realigning classifications on the basis of truth promotes accurate, and that is to say culturally respectful and attentive, mechanisms to classify in ways that align with a variety of true beliefs. In this vein, we must acknowledge that implicit in this statement is the reality that there must be many ways in which the world can be seen as epistemically true. This does not mean, of course, that there are "alternative facts." To say that we should prioritize truth is to also say that, as information specialists, we have a responsibility to say when someone is incorrect or proselytizing information that is unsupported by valid forms of evidence (however we define evidence in a particular context—even ones that are, potentially, antithetical to a Western epistemic notion of science). From an indigenous point of view, truth may be tantamount to a system that promotes balance between the social and natural world, for example, rather than one based on exploitation and nature-as-capital. Failing to attend to alternate views of justice within classification leads to community-specific modes of oppression.

Representational biodiversity systems that do not support underrepresented ways of knowing also facilitate environmental injustices. As defined by the United States Environmental Protection Agency, to be environmentally just means to support

> the fair treatment and meaningful involvement of all people regardless of race, ethnicity, income, national origin or educational level with respect to the development, implementation, and enforcement of environmental laws, regulations, and policies. Fair treatment means that no population, due to policy or economic disempowerment, is forced to bear a disproportionate burden of the negative human health or environmental impacts of pollution or other environmental consequences resulting from industrial, municipal, and commercial operations or the execution of federal, state, local and tribal programs and policies. (Institute of Medicine (US) and Health Sciences Policy Program (US) 1999, 1)

The connection between people and the environment goes well beyond a theoretical analysis of Alfred Whitehead, process theory, or socionatural relations. The relationship is also practical. Within the information disciplines, especially those allied with science and technology and media studies, we understand how technological waste, for example, has adverse effects on rural, marginalized, and underdeveloped communities (Parks and Starosielski 2015; Iles 2004; Kellogg and Mathur 2003; Robyn 2002). Our physical, human bodies are part of the environment, and to think otherwise is to fail to recognize the drastic impacts environmental policies and regulations have on our sense of well-being and health. On top of the material connections between the environment and humanity, we must also acknowledge religious and cultural impacts. The dingo, for example, plays prominently in Australian indigenous creation myth. The dingo plays a central role in spiritual stories associated with The Dreaming, for example, which narrates the unfolding origin story of humanity that, in some reports, is born of the dingo figure (Smith and Litchfield 2009)—a metaphor for our current treatment of the dingo in the most egregious of ways.

Gordon Walker identifies four pillars of a critical environmental justice approach: (1) recognition that social inequality and oppression intersect in all human and nonhuman forms; (2) understanding that injustices happen at different scales—at the local and global levels, for example—as well as how the past influences the material present and how the present redefines the past; (3) recognizing that social inequities are systemic and reinforced by state powers; and (4) acknowledging that *all* bodies—human and nonhuman—are indispensable to our collective well-being and that we must fight against the dominant hegemonic forces of racism, colonialization, and other forces of social suppression (2012, 1). "Socioecological indispensability" is the broad term Walker uses to indicate the importance of *all* natural entities (human and nonhuman) as vital actors in the pursuit of all forms of justice, including social and economic. Environmental justice is a global issue and, as such, requires that information sources take into account multiple forms of global opinion (Walker 2012, chap. 8)—ostensibly what consensus classification might be seen as setting out to do, whether ideally or counter to professional preference.

Which leads to the second notion of justice that also must be attended to: a definition of justice that includes not only the harms performed to human actors, but those to nonhuman actors as well. If we are to say that classifications should facilitate an ethic of *truth*, then that truth should also be one that aligns with a nonanthropocentric, environmentally favorable view of the world. This approach requires us to also subscribe to ecological justice frameworks and be attentive to how classifications can potentially unfavorably burden ecological communities with or without human involvement. Discourses surrounding environmental justice are anthropocentric and focus on groups or communities of individuals that are sentient and can engage in social agreements about what is just in a given particular context (Garner 2013, 4). Garner notes that John Rawls, for example, failed to include animals in his framework because they lacked moral agency, or what Rawls called "the capacity for moral personality" (1999, 443). Rawls's notion of justice is based on the notion of a social contract, which specifically excludes the realm of nature: "This wider theory fails to embrace all moral relationships, since it would seem to include only our relations with other persons and to leave out of account how we are to conduct ourselves toward animals and the rest of nature" (Rawls 1999, 15). Of course, although nonhuman natural organisms may have no ability (that we know of) to internalize or negotiate their own identities in systems, we do have a sense of responsibility to do so on their behalf.

Ecological justice is an approach that attempts to bring into discourses of justice a notion that nature, in and of itself, is a vital part of this social assemblage. David Schlosberg, incorporating the scholarship of Brian Baxter (2014), notes that one can imagine bridging environmental justice with the ecological worldview if we extend the notion of "community" and "claim that viable populations of merely living organisms [non-sentient organisms] have a right to environmental resources necessary for these populations to exist and survive" (Schlosberg 2007, chap. 5). In this way, the stakes for biodiversity and the consequences of its management are distributed (equally or otherwise) throughout all communities of living beings, regardless of whether one has entered into a contract that defines what is or is not just. In such a view, we have a responsibility to consider nonhuman species as part of the decisions we make that equally affect their constitution, in both the present and the future. A community-centric approach also acknowledges that there are, in fact, a good deal of similarities between human and nonhuman organisms, and that such an approach allows us to recognize and empathize with the stakes for the nonhuman natural world (Schlosberg 2007, 133–134). For one, we need the same resources to survive—air, water, land, and the like—and that alone should extend the concept of justice beyond the realm of the human. As aptly stated by Martha C. Nussbaum, "Animals, like humans, pursue a plurality of distinct goods: friendship and affiliation, freedom from pain, mobility, and many others" (2007, 344).

Nature also has a sense of agency in that, when something is awry, we are well aware due to the expression of some systemic disruption. As a result of climate change, for example, nature responds appropriately by expressing itself through redistribution (or extinction) of species. And surely, whether we act as if it is the case or not, humans are not immune to such extinction events. Humans and nonhuman organisms exist within the same system and, to flourish, both aspects of nature must survive in tandem and each deserves "to enjoy the fulness of its own form of life" (Schlosberg 2007, 136–138). As discussed, biodiversity taxonomic work, at its heart, should be seen as a social-ecological concept, and as such, the natural world should be considered when understanding the effects these classifications have on the regulation and control of the natural world.

However, the limitations of consensus structures are apparent when we begin to think about them as tools used in the forum of environmental and ecological justice. The access systems are not yet mature enough to convey the nuances of their arrangement and their potential limitations in the domains of public practice. Common notions of environmental justice rely on a relatively straightforward critique of injustice through the distribution of environmental harms, or on the justice associated with the beneficial aspects of a clean environment (Gilio-Whitaker 2019, 23). And indeed, As Dina Gilio-Whitaker notes, "for a conception of environmental justice to be relevant to a group of people, it must fit within conceptual boundaries that are meaningful to them" (2019, 24). For one, it isn't enough to merely distribute the ill effects of environmental destruction, we must seek an environmental and ecological justice that stops environmental harms altogether. For American indigenous populations, colonialism proper began as an environmental injustice, and thus, any attempt to offset these injustices must reconcile with this settler-colonial past (Gilio-Whitaker 2019, 12). Tied as it is to colonial power, an indigenous environmental justice is one that sees as its goal the repatriation of native land, as well as the deconstruction of the systems that monitor and maintain the capitalist and white power structures. Given that indigenous ways of thinking are intricately tied to the environmental domain, all damage and violence performed to indigenous populations involve, in some way, environmental injustices. This fact taints most, if not all, legal attempts to offset environmental injustices, given that they are rooted, to begin with, in imperialist motivations (Gilio-Whitaker 2019, 149)

To reorient a biodiversity taxonomic framework in such a way means reorienting the ways in which we collect, form, and deliver the *form* of knowledge within classification systems. This is no easy task, but one that has been tackled by some from within the sciences (Alexander et al. 2011; Aswani and Lauer 2014; Ban et al. 2018; Reyes-García et al. 2016). Without a more promiscuous-realist and pluralist conception of classification—a collective biodiversity system in which, at least, *some* classifications represent indigenous ontological and epistemological commitments—efforts to overcome ecological and environmental injustices in the eyes of indigenous populations will fail from the start. We cannot begin to liberate nature on indigenous terms if the primary tool to facilitate this liberation is based on the scientific colonial powers that imprisoned it in the first place. The difficulty in incorporating indigenous ways of understanding nature into taxonomic work arises, at least in part, from the ingrained belief in the nature-human bifurcation: indigenous culture's relation to nature, and the scientific knowledge that proceeds from it, never acknowledged such a distinction. Looking again to Gilio-Whitaker (2019, 27), "the very thing that distinguishes Indigenous peoples from settler societies is their unbroken connection to ancestral lands." What we have broken, nonetheless, we must continually try to mend.

CONCLUSION: POSITION IS POWER

Looping back to my opening narrative about the Global Taxonomy Initiative established by the Convention on Biological Diversity in 1998: Nearly 25 years later, even with the impressive advancements in all facets of biodiversity science—from DNA barcoding to large scale computational phylogenetics—the field still has some of its greatest challenges in front of it. Even with the problems associated with managing data aggregation, the limitations and problematics of algorithmic software, and need for high-speed computing for speculative evolutionary purposes, the greatest challenges for biodiversity science, it seems to me, are social and cultural more than they are technological. On the whole, biodiversity science is not a computationally intensive endeavor, at least not when juxtaposed with other largescale collaborative scientific initiatives such as CERN, which requires a $5 billion Large Hadron Collider, in addition to a bevy of other accelerators and decelerators, to perform a core function of its work. The workhorse of the biodiversity is the relational database—one of the most integral and overlooked tools on our technological and scientific landscape. The core challenges, as noted by the Alliance for Biodiversity Knowledge (Hobern et al. 2019), are in part the need to coordinate global skills and activities, develop shared roadmaps, and to manage shared and collective governance. These are human problems, problems of social and collective epistemology. The increasingly international and collaborative nature of biodiversity research is requiring a shift in consciousness in the field that prioritizes data sharing, open science, and the emergence of standards that, to at least a certain extent, prioritize data communication for the benefit of the greater scientific good. And at the heart of this cooperation are biological classifications that, by virtue of their communicative importance, have emerged as central spaces of promise and of conflict.

An aim of this narrative has been to expose the ways that classifications in general, and biodiversity classifications in specific, create, expose, distribute, and sustain various modes of power. This power is representational, it is material, it is epistemic, and fundamental. If we recall our hypothetical WeevilBase from chapter 3, we can imagine the numerous possible online repositories this data has populated by way of structures such as the Catalogue of Life and GBIF. Scientists from every nation the world could have downloaded WeevilBase data and implemented it in any number of ways. One problem is that we may know how many times WeevilBase was accessed or downloaded (by way of platform metrics), but we cannot for certain know for what purposes or where the data potentially flowed thereafter. This data, now activated in these disparate spaces, has the capacity to inform a host of environmental and ecological decisions and interventions. But like so much information, the further it gets from its context, the less control creators have over its interpretation and production context.

The Catalogue of Life is an instructive infrastructure because, while it facilitates the necessary sharing of information, it also pushes against many of the prevailing assumptions that have been central to the practice of biodiversity taxonomy for hundreds of years, particularly that the primary role of biological taxonomies is to present a formal and unified presentation of an argued hypothesis. As a result of this turn toward access, a system like the Catalogue shows what we lose epistemically when we design systems primarily for broad and varied uses. Local knowledge and context are lost to the benefit of global data coordination and control. Perhaps most importantly, the mixed reception and critiques of composite systems starkly expose the fact that classifications have epistemic impacts *at all*.

The classification issues discussed throughout this text are, I believe, central to the production of human culture in our current epistemic environments—even beyond the narrow limits of environmental culture and biodiversity work. We know that platforms like the Catalogue of Life are shaping the outline of biodiversity knowledge in educational environments. The Encyclopedia of Life (2018), for example, offers a number of lesson plans for grades 2 to 12, and by virtue of these (admittedly, useful and indispensable) resources, the logic of consensus classifications is now becoming the default for many students just as textbook models of the classification of dinosaurs dominated my own youth. But more significantly, biodiversity classifications deeply affect the way humanity imagines itself to live in, and in relation to, this natural world. What biodiversity taxonomists take for granted—that classifications are artificial and constructed, that they are arguments and subject to fruitful disagreement—is just not an acknowledged or normalized reality in our public consciousness. Just as individuals depend on the seemingly objective quality of search engines for their information retrieval purposes, so too do individuals see trees of life and other graphical representations of the natural world as static and singularly authoritative. If any biodiversity taxonomist thinks otherwise, I urge you to please spend a few days at the reference desk of a library (as I did for many years); I can assure you, the first impulse for the general user is not to question the veracity of most of the information they receive. Our collective, public ability to interrogate our informational world is staggeringly lacking.

So, when structures like the Catalogue are constructed with layers of complex epistemic entanglements, we need to be aware of how best to facilitate systemic transparency so that individuals can appropriately contextualize this information. It is imperative that academics in all disciplines (scientists, humanists, artists, computer science, and all) work together in ways that can make the internal schematics of information structures more visible—to foster a culture of questioning with regard to how we navigate technologies and to ingrain a sense of critical inquiry and an ethic bent toward openness and justice. The more concertedly we do this, the more sensitive our collective publics will be to the reality of what these contingent and opaque structures illustrate—and what they do not. Public trust in science and authority is at an all-time low, and although it's impossible to pinpoint any one reason why this is the case (surely there are many), we can certainly say that the information infrastructures we create are not helping to ameliorate this trend. Of course, changing the course of the public perception of information is not the sole responsibility of the Catalogue—or any one system, for that matter. The point here is that the Catalogue's structure and composition does give us the opportunity to ask productive questions about the future methodologies of applications of classification and taxonomic work.

Scholars in information studies have worked for decades to make finding information easy, to optimize our user environments for ease and comfort, and to deliver relevant information. But, alas, the pendulum has swung too far in one direction: it is now too easy to find the information we think we need and far more difficult to find information that is useful and true. This must change, and for this to happen, we have to feel comfortable critiquing the very foundation of our information practices—even at the expense of slowing our efforts and working through some of the disciplinary problems these efforts expose.

With this said, it is important to note that classifications are not only the foundation for our production of knowledge and understanding of the natural world; equally impactful are their material consequences on the lives of countless species around the globe. Throughout this text I've discussed a number of hypothetical organisms as exemplars for my discussions—dingoes, beetles, bears, parrots, and mussels, to name a few. The examples may be hypothetical, but the circumstances related to the peril of these species are not. It is relatively easy to perhaps turn one's head and say, *Well, this is the natural course of things!* But nothing about the climate, extinction rates, the fires, the flooding, and the prevalence of disease is happenstance. It is our human inclination to exploit that produced these natural ills, and this is a problem.

At this moment, I am in my living room at what I hope will be, in retrospect, the beginning of the long end to the COVID-19 pandemic. National and international discourse is such that economic hardships take precedence over the lives that have been lost and on the toll this has taken on our collective global psyche. Meanwhile, the environmental circumstances that exacerbate these problems continue to go undiscussed, if not wholly disregarded. As noted by the Harvard Chan Center for Climate,

Health, and the Global Environment (C-CHANGE), "Many of the root causes of climate change also increase the risk of pandemics. Deforestation, which occurs mostly for agricultural purposes, is the largest cause of habitat loss worldwide. Loss of habitat forces animals to migrate and potentially contact other animals or people and share germs. Large livestock farms can also serve as a source for spillover of infections from animals to people" (2020). This imbalance and disregard for our natural surrounding is our proverbial canary in the coal mine. This scenario, and many others like it, raises the stakes for work in the biodiversity taxonomic sciences. It is, in part, because of biodiversity data produced over the course of hundreds of years that we have a baseline to understand the rate and impact of our ecological and environmental change.

And so, with the stakes as high as they are, we can better understand why the material and aesthetic qualities of classifications have been so important to the narrative of this text. Our position in a classification—our derivative positionality—is important because it serves as the nexus between our affective experience in the world and the abstract representational space that is so often easy to overlook. But that representational space is integral to how we function within society—and more importantly, how we think about our own identity in relation to human and nonhuman entities. If *designers* of systems are not giving due attention to the diverse potentials of these spaces, then we will never reclaim our technological identities, nor the power that they have over us in our daily interactions. We have to continually argue that classification space *is* justice space, and that the lived world will continue to be unjust unless our derivative identities reflect an alternate reality based on equality and equity. It is important for IS scholars and practitioners to realize that they have a role to play in how these online infrastructures work in service to these ends.

But even beyond personal identities, it is important for all information specialists to invest in this broader project of systemic deconstruction and critique for those who cannot act on their own behalf. I am not speaking merely of the human world, but also of the natural world. The species on this planet have everything to lose but have absolutely no agency with regard to our decisions and actions. Social justice *is also* environmental and ecological justice; one cannot exist without the others. To speak disciplinarily: information studies would do well to examine new modes of justice beyond those that apply to anthropocentric concerns—they can, perhaps, help us reenvision new potentials for classificatory arrangements. The analytic I illustrate in figure 0.1 is only a small part of the larger critique of power as it relates to representational and informational spaces. In our social reality, our social (economic, education, etc.) position dictates the capacities of our social powers, while in our represented spaces, our power is relative to our classified position. The fate of the dingo is one thing, but the reality is that our human fate is as contingent and dependent on these systems as any other organism. Our social and representational positions are fragile, so we should tend to them with the care and capacities our physical selves deserve.

Those that work within the broad discipline of information studies have an obligation to the public to view spaces of classification as facilitating both epistemic empowerment and epistemic injustices. And while we may not be able to define any *exact* notion of how we define empowerment or justice universally (and it would be dangerous to do so, it seems to me), we certainly can and should design our interventions to create systems that accept many sources and structures of knowledge. We must *try*, even if the end result is imperfect on the pathway to some semblance of success. In any geographic and political space, we will be materially, politically, economically, and socially challenged in some way. But we should still continue to acknowledge that if we start dialogue; if we enact critical literacies; if we design and implement our services, programs, and collections in culturally diverse ways, then we have, properly, to the best of our ability, set the foundation for whatever change is necessary—whether it is *our* concept of change or not. Using humanistic approaches based on historical, philosophical, and cultural contexts is the best that we can do. And surely, if there was a time to do so, that time is certainly the present. All said, if I—if we—are to succeed in these aims, then surely libraries and the information disciplines will be an integral part of this change. But we must first acknowledge that any change in the world will, at some point, contend with the activity of classification—it is, after all, one of our most basic human instruments. Such is the impact of space represented: position is power.

Notes

CHAPTER 3

1. The goal of 2.3 million species is articulated by the Catalogue of Life, though numbers differ widely as to how many species actually exist on the planet (Eng 2016; Hug et al. 2016; Zimmer 2016). David Hill's Open Tree of Life Project projects 2.3 million species (Hinchliff et al. 2015; "'Tree of Life' for 2.3 Million Species Released" 2015), while a recent study estimated over one trillion microbial species alone on the planet (Locey and Lennon 2016). For the sake of consistency, when referenced in this book, the total number of species on the planet will be assumed to be approximately 2.3 million.

CHAPTER 4

1. See also International Association for Plant Taxonomy 2011, sec. 2.8 for international code of nomenclature for algae, fungi, and plants, which takes a similar approach.
2. Where the italicized name is the valid scientific name (genus and species) that a scientist, Snyder, described and published in 1904. *Sensu* (often abbreviated *sec*) is a Latin term that means "in the sense of" and is often used at the end of names to indicate that the author used the species concept "in the sense of" whoever is cited. Thus, since Snyder described the species for the first time in 1904, it is *also* Snyder concept.
3. Where ">" means "the synonym of," and "+" indicates the two species concepts melding together to form one larger concept (in this case, that of *Centropyge fisheri* (Snyder 1904) sec. Pyle 2003).

CHAPTER 5

1. I use ontology here in the computational sense, in terms of modeling the world, rather than the philosophical sense, describing fundamental categories of things.
2. See also Hennig, Davis, and Zangerl 1999.

References

Adler, Melissa. 2017. *Cruising the Library: Perversities in the Organization of Knowledge*. New York: Fordham University Press.

Aguilera, Jasmine. 2016. "Another Word for 'Illegal Alien' at the Library of Congress: Contentious." *New York Times*, July 22. https://www.nytimes.com/2016/07/23/us/another-word-for-illegal-alien-at-the-library-of-congress-contentious.html.

Alexander, Clarence, Nora Bynum, Elizabeth Johnson, Ursula King, Tero Mustonen, Peter Neofotis, Noel Oettlé, et al. 2011. "Linking Indigenous and Scientific Knowledge of Climate Change." *BioScience* 61 (6): 477–484. https://doi.org/10.1525/bio.2011.61.6.10.

Allen, L., and P. Fleming. 2004. "Review of Canid Management in Australia for the Protection of Livestock and Wildlife—Potential Application to Coyote Management." *Sheep & Goat Research Journal*, October. https://digitalcommons.unl.edu/icwdmsheepgoat/2.

Allen, L. R., and E. C. Sparkes. 2001. "The Effect of Dingo Control on Sheep and Beef Cattle in Queensland." *Journal of Applied Ecology* 38 (1): 76–87. https://doi.org/10.1046/j.1365-2664.2001.00569.x.

Andersen, Jack, and Laura Skouvig, eds. 2017. *The Organization of Knowledge: Caught between Global Structures and Local Meaning*. Studies in Information. Bingley, UK: Emerald Publishing.

Anderson, Elizabeth. 2020. "Feminist Epistemology and Philosophy of Science." In *The Stanford Encyclopedia of Philosophy*, edited by Edward N. Zalta. Stanford, CA: Stanford University Metaphysics Research Lab. https://plato.stanford.edu/archives/spr2020/entries/feminism-epistemology/.

Anderson, Robert, Miguel Araújo, Antoine Guisan, Jorge M. Lobo, Enrique Martínez-Meyer, A. Townsend Peterson, and Jorge Soberón. 2016. "Report of the Task Group on GBIF Data Fitness for Use in Distribution Modelling." Text. gbif.org. March 22. http://www.gbif.org/resource/82612.

Archibald, Jo-Ann, Jenny Lee-Morgan, and Jason De Santolo. 2019. *Decolonizing Research: Indigenous Storywork as Methodology*. London: Zed Books.

Arias-Maldonado, Manuel. 2015. *Environment and Society: Socionatural Relations in the Anthropocene.* https://doi.org/10.1007/978-3-319-15952-2.

Aristoteles. 1995. *The Complete Works of Aristotle. Vol. 1: [. . .].* Edited by Jonathan Barnes. 6th printing, Bollingen Series 71-2. Princeton, NJ: Princeton University Press.

Arnellos, Argyris. 2018. "From Organizations of Processes to Organisms and Other Biological Individuals." In *Everything Flows: Towards a Processual Philosophy of Biology,* 1st ed., edited by Daniel J. Nicholson and John Dupré. Oxford: Oxford University Press.

Aswani, Shankar, and Matthew Lauer. 2014. "Indigenous People's Detection of Rapid Ecological Change." *Conservation Biology* 28 (3): 820–828. https://doi.org/10.1111/cobi.12250.

Ballard, J. William O., and Laura A. B. Wilson. 2019. "The Australian Dingo: Untamed or Feral?" *Frontiers in Zoology* 16 (1): 2. https://doi.org/10.1186/s12983-019-0300-6.

Bamford, Matt. 2018. "Dingoes to Remain Classified as Non-Native Wild Dogs under WA Law Reform." ABC Australia. August 28. https://www.abc.net.au/news/2018-08-28/dingoes-will-no-longer-be-native-animals-in-western-australia/10172448.

Ban, Natalie C., Alejandro Frid, Mike Reid, Barry Edgar, Danielle Shaw, and Peter Siwallace. 2018. "Incorporate Indigenous Perspectives for Impactful Research and Effective Management." *Nature Ecology & Evolution* 2 (11): 1680–1683. https://doi.org/10.1038/s41559-018-0706-0.

Barr, Catherine, and Anne Wilson. 2018. *Red Alert! Endangered Animals around the World.* 1st US ed. Watertown, MA: Charlesbridge.

Baxter, Brian. 2014. *A Theory of Ecological Justice.* 1st ed. London: Routledge.

Beghtol, Clare. 1986. "Semantic Validity: Concepts of Warrant in Bibliographic Classification Systems." *Library Resources & Technical Services* 30:109–125.

Beghtol, Clare. 2001. "Relationships in Classificatory Structures and Meaning." In Relationships in the Organization of Knowledge, edited by Carol A. Bean and Rebecca Green, 99–113. Boston: Kluwer Academic Publishers.

Bellard, Céline, Cleo Bertelsmeier, Paul Leadley, Wilfried Thuiller, and Franck Courchamp. 2012. "Impacts of Climate Change on the Future of Biodiversity." *Ecology Letters* 15 (4): 365–377. https://doi.org/10.1111/j.1461-0248.2011.01736.x.

Bennett, Jane. 2010. *Vibrant Matter: A Political Ecology of Things.* Durham, NC: Duke University Press.

Bercovitch, Fred B., Philip S. M. Berry, Anne Dagg, Francois Deacon, John B. Doherty, Derek E. Lee, Frédéric Mineur, et al. 2017. "How Many Species of Giraffe Are There?" *Current Biology* 27 (4): R136–R137. https://doi.org/10.1016/j.cub.2016.12.039.

Bisby, Frank A. 2000. "The Quiet Revolution: Biodiversity Informatics and the Internet." *Science* 289 (5488): 2309–2312. https://doi.org/10.1126/science.289.5488.2309.

Bisby, Frank A., Junko Shimura, Michael Ruggiero, James Edwards, and Christoph Haeuser. 2002. "Taxonomy, at the Click of a Mouse." *Nature* 418 (6896): 367–367. https://doi.org/10.1038/418367a.

Bliss, Henry Evelyn. 1929. *The Organization of Knowledge and the System of the Sciences.* New York: H. Holt and Company. http://catalog.hathitrust.org/Record/001388383.

Bliss, Henry Evelyn. 1933. *The Organization of Knowledge in Libraries and the Subject-Approach to Books.* New York: H.W. Wilson Company.

Bock, Walter J. 2004. "Species: The Concept, Category and Taxon." *Journal of Zoological Systematics and Evolutionary Research* 42 (3): 178–190. https://doi.org/10.1111/j.1439-0469.2004.00276.x.

Bone, Christine, and Brett Lougheed. 2018. "Library of Congress Subject Headings Related to Indigenous Peoples: Changing LCSH for Use in a Canadian Archival Context." *Cataloging & Classification Quarterly* 56 (1): 83–95. https://doi.org/10.1080/01639374.2017.1382641.

Borges, Jorge Luis. 1999. *Selected Non-Fictions.* Edited by Eliot Weinberger. New York: Viking.

Bourgoin, Thierry. 2016. "Importance of Databasing Taxonomic Knowledge: An Example with FLOW (Fulgoromorpha Lists on the Web). (Invited Communication)." November 14. Beijing: Beijing Forestry University.

Bowers, Fredson. 1994. *Principles of Bibliographical Description.* St. Paul's Bibliographies 15. Winchester, UK: Oak Knoll Press.

Bowker, Geoffrey C. 2008. *Memory Practices in the Sciences.* Cambridge, MA: MIT Press.

Bowker, Geoffrey C., and Susan Leigh Star. 1999. *Sorting Things Out: Classification and Its Consequences.* Inside Technology. Cambridge, MA: MIT Press.

Briet, Suzanne. 1951. *What Is Documentation? English Translation of the Classic French Text.* Translated by Ronald E. Day. Lanham, MD: Scarecrow Press.

Broadfield, A. 1946. *The Philosophy of Classification.* 1st ed. London: Grafton & Co.

Buckland, Michael. 2017. "Reflections on Suzanne Briet." In Proceedings of the 11th French ISKO Colloquium. Paris: ISTE Editions.

Burke, Anthony, and Stefanie Fishel. 2019. "Power, World Politics, and Thing-Systems in the Anthropocene." In *Anthropocene Encounters: New Directions in Green Political Thinking,* edited by Frank Biermann and Eva Lövbrand. Cambridge: Cambridge University Press.

Butchart, Stuart H. M., Matt Walpole, Ben Collen, Arco van Strien, Jörn P. W. Scharlemann, Rosamunde E. A. Almond, Jonathan E. M. Baillie, et al. 2010. "Global Biodiversity: Indicators of Recent Declines." *Science* 328 (5982): 1164–1168. https://doi.org/10.1126/science.1187512.

Butler, Chris. 2014. *Henri Lefebvre: Spatial Politics, Everyday Life and the Right to the City.* Nomikoi: Critical Legal Thinkers. Abingdon, UK: Routledge.

Butler, Judith. 2000. *Contingency, Hegemony, Universality.* London: Verso.

Butler, Judith. 2020. *The Force of Nonviolence: An Ethico-Political Bind.* Brooklyn: Verso Books.

Cachuela-Palacio, Monalisa. 2006. "Towards an Index of All Known Species: The Catalogue of Life, Its Rationale, Design and Use." *Integrative Zoology* 1 (1): 18–21. https://doi.org/10.1111/j.1749-4877.2006.00007.x.

Cairns, Kylie M., Katherine Moseby, Euan Ritchie, Rob Appleby, Peter Savolainen, Arian Wallach, and Melanie Fillios. 2020. "South Australia Is Still Killing Dingoes." *ConservationBytes.Com* (blog). April 14. https://conservationbytes.com/2020/04/14/south-australia-is-still-killing-dingoes/.

CatalogueOfLife/ChecklistBank. 2021 [2017]. JavaScript. CoL. https://github.com/CatalogueOfLife/checklistbank.

Cetina, Karin Knorr. 1999. *Epistemic Cultures: How the Sciences Make Knowledge*. Cambridge, MA: Harvard University Press.

Chambers, Lynda E., Phoebe Barnard, Elvira S. Poloczanska, Alistair J. Hobday, Marie R. Keatley, Nicky Allsopp, and Les G. Underhill. 2017. "Southern Hemisphere Biodiversity and Global Change: Data Gaps and Strategies." *Austral Ecology* 42 (1): 20–30. https://doi.org/10.1111/aec.12391.

Chen, Yi-Yun, Nico Franz, Jodi Schneider, Shizhuo Yu, Thomas Rodenhausen, and Bertram Ludäscher. 2017. "Agreeing to Disagree: Reconciling Conflicting Taxonomic Views Using a Logic-Based Approach." *Proceedings of the Association for Information Science and Technology* 54 (1): 46–56. https://doi.org/10.1002/pra2.2017.14505401006.

Conservation International. 2017. "Why Hotspots Matter." Conservation International. https://www.conservation.org/priorities/biodiversity-hotspots.

Convention on Biological Diversity. 2003. "Guide to the Global Taxonomic Initiative." Secretariat of the Convention on Biological Diversity. https://www.cbd.int/doc/publications/cbd-ts-30.pdf.

Convention on Biological Diversity. 2006. "Rio Declaration on Environment and Development." Convention on Biological Diversity. Secretariat of the Convention on Biological Diversity. November 13. https://www.cbd.int/doc/ref/rio-declaration.shtml.

Convention on Biological Diversity. 2016. "History of the Convention." https://www.cbd.int/history/default.shtml.

Convention on Biological Diversity. 2017a. "Global Taxonomy Initiative: Background." https://www.cbd.int/gti/background.shtml.

Convention on Biological Diversity. 2017b. "The Convention on Biological Diversity." https://www.cbd.int/convention/.

"Convention on Biological Diversity (Full Text)." 1992. United Nations. https://www.cbd.int/doc/legal/cbd-en.pdf.

Corn, Aaron, with Steven Wantarri Jampijinpa Patrick. 2019. "Exploring the Applicability of the Semantic Web for Discovering and Navigating Australian Indigenous Knowledge Resources." *Archives and Manuscripts* 47 (1): 131–152. https://doi.org/10.1080/01576895.2019.1575248.

Critical Ecosystem Partnership Fund. 2017. "Explore the Biodiversity Hotspots | CEPF." Critical Ecosystem Partnership Fund. https://www.cepf.net/our-work/biodiversity-hotspots.

Croft, J., N. Cross, S. Hinchcliffe, E. Nic Lughadha, P. F. Stevens, J. G. West, and G. Whitbread. 1999. "Plant Names for the 21st Century: The International Plant Names Index, a Distributed Data Source of General Accessibility." *Taxon* 48 (2): 317–324. https://doi.org/10.2307/1224436.

Culham, A., and C. Yessen. 2018. "Droseraceae Database (Version 0.1, Dec 2008)." In *Species 2000 & ITIS Catalogue of Life*. January 30. Leiden: Species 2000: Naturalis. http://www.catalogueoflife.org/col/details/database/id/66.

Dahlberg, Ingetraut. 2017. "Brief Communication: Why a New Universal Classification System Is Needed." *Knowledge Organization* 44 (1): 65–71.

Danton, J. Periam. 1973. *The Dimensions of Comparative Librarianship*. Chicago: American Library Association.

Daston, Lorraine. 2004. "Type Specimens and Scientific Memory." *Critical Inquiry* 31 (1): 153–182. https://doi.org/10.1086/427306.

Day, Ronald E. 2011. "Death of the User: Reconceptualizing Subjects, Objects, and Their Relations." *Journal of the American Society for Information Science and Technology* 62 (1): 78–88. https://doi.org/10.1002/asi.21422.

Day, Ronald E. 2014. *Indexing It All: The Subject in the Age of Documentation, Information, and Data*. 1st ed. Cambridge, MA: MIT Press.

De Queiroz, Kevin. 2007. "Species Concepts and Species Delimitation." *Systematic Biology* 56 (6): 879–886. https://doi.org/10.1080/10635150701701083.

Dickinson, Emily, and Thomas Herbert Johnson. 1997. *The Complete Poems of Emily Dickinson*. Boston: Back Bay Books.

Doolittle, W. Ford. 1999a. "Phylogenetic Classification and the Universal Tree." *Science* 284 (5423): 2124–2128. https://doi.org/10.1126/science.284.5423.2124.

Doolittle, W. Ford. 1999b. "Phylogenetic Classification and the Universal Tree." *Science* 284 (5423): 2124–2128. https://doi.org/10.1126/science.284.5423.2124.

Döring, Markus. 2015. "Developer Blog: Improving the GBIF Backbone Matching." *Developer Blog* (blog). March 30. http://gbif.blogspot.com/2015/03/improving-gbif-backbone-matching.html.

Drescher, Jack. 2015. "Out of DSM: Depathologizing Homosexuality." *Behavioral Sciences* 5 (4): 565–575. https://doi.org/10.3390/bs5040565.

Drucker, Johanna. 2014a. "Distributed and Conditional Documents: Conceptualizing Bibliographical Alterities." *Materialities of Literature* 2 (1): 11–29. https://doi.org/10.14195/2182-8830_2-1_1.

Drucker, Johanna. 2014b. *Graphesis: Visual Forms of Knowledge Production*. Cambridge, MA: Harvard University Press.

Duarte, Marisa Elena. 2017. *Network Sovereignty: Building the Internet across Indian Country*. Indigenous Confluences. Seattle: University of Washington Press.

Duarte, Marisa Elena, and Miranda Belarde-Lewis. 2015. "Imagining: Creating Spaces for Indigenous Ontologies." *Cataloging & Classification Quarterly* 53 (5/6): 677–702. https://doi.org/10.1080/01639374.2015.1018396.

Dupré, John. 1993. *The Disorder of Things: Metaphysical Foundations of the Disunity of Science.* Cambridge, MA: Harvard University Press.

Dupré, John. 1996. *The Disorder of Things: Metaphysical Foundations of the Disunity of Science.* 3rd ed. Cambridge, MA: Harvard University Press.

Dupré, John. 2014. *Processes of Life: Essays in the Philosophy of Biology.* Oxford: Oxford University Press.

Edwards, Paul N. 2010. *A Vast Machine: Computer Models, Climate Data, and the Politics of Global Warming.* Cambridge, MA: MIT Press.

Egan, Margaret E., and Jesse H. Shera. 1952. "Foundations of a Theory of Bibliography." *Library Quarterly* 22 (2): 125–137. https://doi.org/10.1086/617874.

Eliot, T. S., and Mary Karr. 2002. *The Waste Land and Other Writings.* New York: Random House.

Ellis, Erle C. 2018. *Anthropocene: A Very Short Introduction.* 1st ed. Very Short Introductions. Oxford: Oxford University Press.

Encyclopedia of Life. 2017. "Brown Bear, Grizzly Bear—Ursus Arctos—Classifications." Encyclopedia of Life. http://eol.org/pages/328581/names.

Encyclopedia of Life. 2018. "EoL: Education." Encyclopedia of Life. http://eol.org/info/ed_resources.

Eng, Karen. 2016. "A Newly Drawn Tree of Life Reminds Us to Question What We Know." *Ideas.Ted.Com* (blog). April 21. http://ideas.ted.com/a-newly-drawn-tree-of-life-reminds-us-to-question-what-we-know/.

Entwistle, Abigail, and Nigel Dunstone. 2000. *Priorities for the Conservation of Mammalian Diversity: Has the Panda Had Its Day?* Cambridge: Cambridge University Press.

Ereshefsky, Marc. 2001. "Names, Numbers and Indentations: A Guide to Post-Linnaean Taxonomy." *Studies in History and Philosophy of Science Part C: Studies in History and Philosophy of Biological and Biomedical Sciences* 32 (2): 361–383. https://doi.org/10.1016/S1369-8486(01)00004-8.

Ereshefsky, Marc. 2007. *The Poverty of the Linnaean Hierarchy: A Philosophical Study of Biological Taxonomy.* 1st ed. Cambridge: Cambridge University Press.

Erickson, David L., and Amy C. Driskell. 2012. "Construction and Analysis of Phylogenetic Trees Using DNA Barcode Data." In *DNA Barcodes,* edited by W. John Kress and David L. Erickson, 395–408. Totowa, NJ: Humana Press. http://link.springer.com/10.1007/978-1-61779-591-6_19.

Escobar, Arturo. 2018. *Designs for the Pluriverse: Radical Interdependence, Autonomy, and the Making of Worlds.* New Ecologies for the Twenty-First Century. Durham, NC: Duke University Press.

Escobar, Arturo. 2020. *Pluriversal Politics: The Real and the Possible.* Latin America in Translation. Durham, NC: Duke University Press.

Fennessy, Julian, Tobias Bidon, Friederike Reuss, Vikas Kumar, Paul Elkan, Maria A. Nilsson, Melita Vamberger, Uwe Fritz, and Axel Janke. 2016. "Multi-Locus Analyses Reveal Four Giraffe

Species Instead of One." *Current Biology* 26 (18): 2543–2549. https://doi.org/10.1016/j.cub.2016.07.036.

Ferraris, Maurizio, and Richard Davies. 2013. *Documentality: Why It Is Necessary to Leave Traces.* New York: Fordham University Press.

Findlen, Paula. 1994. *Possessing Nature: Museums, Collecting, and Scientific Culture in Early Modern Italy.* Studies on the History of Society and Culture 20. Berkeley: University of California Press.

Fischer, Suzanne. 2012. "Nota Bene: If You 'Discover' Something in an Archive, It's Not a Discovery." *The Atlantic,* June 19. https://www.theatlantic.com/technology/archive/2012/06/nota-bene-if-you-discover-something-in-an-archive-its-not-a-discovery/258538/.

Fishbase. 2017. "FishBase." Fishbase. February. http://www.fishbase.org/search.php.

Foucault, Michel. 1995. *Discipline & Punish: The Birth of the Prison.* Translated by Alan Sheridan. 2nd ed. New York: Vintage Books.

Foucault, Michel. 2007. *The Order of Things: An Archaeology of the Human Sciences.* Repr. Routledge Classics. London: Routledge.

Foucault, Michel, and Colin Gordon. 1980. *Power/Knowledge: Selected Interviews and Other Writings, 1972–1977.* 1st US ed. New York: Pantheon Books.

Franz, N. 2008. "Revision, Phylogeny and Natural History of Cotithene Voss (Coleoptera: Curculionidae)." https://doi.org/10.11646/ZOOTAXA.1782.1.1.

Franz, Nico. 2020. "Taxonbytes (@taxonbytes) / Twitter." Twitter. https://twitter.com/taxonbytes.

Franz, Nico M. 2005. "On the Lack of Good Scientific Reasons for the Growing Phylogeny/Classification Gap." *Cladistics* 21 (5): 495–500. https://doi.org/10.1111/j.1096-0031.2005.00080.x.

Franz, Nico M. 2010. "Biological Taxonomy and Ontology Development: Scope and Limitations." *Biodiversity Informatics* 7 (1). https://doi.org/10.17161/bi.v7i1.3927.

Franz, Nico M., and Beckett W. Sterner. 2018. "To Increase Trust, Change the Social Design behind Aggregated Biodiversity Data." *Database* 2018 (January). https://doi.org/10.1093/database/bax100.

Franz, Nico, Robin K. Peet, and Alan S. Weakley. 2008. "On the Use of Taxonomic Concepts in Support of Biodiversity Research and Taxonomy." In *The New Taxonomy,* edited by Quentin Wheeler. The Systematics Association Special Volume Series 76. Boca Raton, FL: CRC Press.

Fricker, Miranda. 2009. *Epistemic Injustice: Power and the Ethics of Knowing.* 1st paperback ed. Oxford: Oxford University Press.

Fuchs, Christian. 2019. "Henri Lefebvre's Theory of the Production of Space and the Critical Theory of Communication." *Communication Theory* 29 (2): 129–150. https://doi.org/10.1093/ct/qty025.

Fuller, Steve. 2009. "Social Epistemology." In *Encyclopedia of Library and Information Sciences, Third Edition,* edited by Mary Niles Maack and Marcia Bates. Boca Raton, FL: CRC Press. https://doi.org/10.1081/E-ELIS3.

Fuller, Steve. 2007. *The Knowledge Book: Key Concepts in Philosophy, Science and Culture.* Stocksfield, UK: Acumen.

Furner, Jonathan. 2004. "'A Brilliant Mind': Margaret Egan and Social Epistemology." https://www.ideals.illinois.edu/handle/2142/1698.

Furner, Jonathan. 2009a. "Dewey Deracialized: A Critical Race-Theoretic Perspective." *Knowledge Organization* 34 (3): 144–168.

Furner, Jonathan. 2009b. "Interrogating Identity: A Philosophical Approach to an Enduring Issue in Knowledge Organization." *Knowledge Organization* 36 (1): 3–16.

Furner, Jonathan. 2013. "ASIS&T Annual Meeting Pre-Conference Activities: The 23rd Annual SIG/CR Classification Research Workshop: A Report." *Bulletin of the American Society for Information Science and Technology* 39 (3): 28–32. https://doi.org/10.1002/bult.2013.1720390309.

Furner, Jonathan. 2016. "Type-Token Theory and Bibliometrics." In *Theories of Informetrics and Scholarly Communication: A Festschrift in Honor of Blaise Cronin*, edited by Cassidy R. Sugimoto and Blaise Cronin. Berlin: De Gruyter.

Furner, Jonathan. 2018. "Truth, Relevance, and Justice: Towards a Veritistic Turn for KO." *Advances in Classification Research* 16:468–474.

Garner, Robert. 2013. *A Theory of Justice for Animals: Animal Rights in a Nonideal World.* New York: Oxford University Press.

Garraffoni, André Rinaldo Senna, and André Victor Lucci Freitas. 2017. "Photos Belong in the Taxonomic Code." *Science* 355 (6327): 805. https://doi.org/10.1126/science.aam7686.

Gaskell, Philip. 2007. *A New Introduction to Bibliography*. Reprinted with corrections in 1995. New Castle, DE: Oak Knoll Press.

GBIF (Global Biodiversity and Information Facility). 2016. "Data Processing." Text. gbif.org. February 25. http://www.gbif.org/infrastructure/processing.

GBIF (Global Biodiversity and Information Facility). 2017a. "Gbif/Checklistbank." GitHub. https://github.com/gbif/checklistbank.

GBIF (Global Biodiversity and Information Facility). 2017b. "Mdoering/Backbone." GitHub. https://github.com/mdoering/backbone.

GBIF (Global Biodiversity and Information Facility). 2017c. "GBIF." http://www.gbif.org/.

GBIF (Global Biodiversity and Information Facility). 2017d. "GBIF Backbone Taxonomy—Constituents." http://www.gbif.org/dataset/d7dddbf4-2cf0-4f39-9b2a-bb099caae36c/constituents.

GBIF (Global Biodiversity and Information Facility). 2019. "An Alliance for Biodiversity Knowledge." https://www.biodiversityinformatics.org/.

GBIF (Global Biodiversity and Information Facility). 2020a. "What Is GBIF?" Text. gbif.org. http://www.gbif.org/what-is-gbif.

GBIF (Global Biodiversity and Information Facility). 2020b. "The GBIF Secretariat." gbif.org. https://www.gbif.org/publisher/fbca90e3-8aed-48b1-84e3-369afbd000ce.

Geniusz, Wendy Djinn. 2009. *Our Knowledge Is Not Primitive: Decolonizing Botanical Anishinaabe Teachings*. 1st ed. The Iroquois and Their Neighbors Series. Syracuse, NY: Syracuse University Press.

Gewin, Virginia. 2002. "Taxonomy: All Living Things, Online." *Nature* 418 (6896): 362–363. https://doi.org/10.1038/418362a.

Gilio-Whitaker, Dina. 2019. *As Long as Grass Grows: The Indigenous Fight for Environmental Justice from Colonization to Standing Rock*. Boston: Beacon Press.

GNA (Global Names Architecture). 2020a. "Global Names Architecture: Glossary." http://globalnames.org/docs/glossary/.

GNA (Global Names Architecture). 2020b. "Good and Bad Names." http://globalnames.org/docs/good-bad-names/.

GNA (Global Names Architecture). 2021a. "Global Names Index." http://gni.globalnames.org/.

GNA (Global Names Architecture). 2021b. "Global Names Resolver." http://resolver.globalnames.org/.

GNA (Global Names Architecture). 2021c. "GlobalNames Home." http://globalnames.org/.

Gnoli, Claudio, and Riccardo Ridi. 2014. "Unified Theory of Information, Hypertextuality and Levels of Reality." *Journal of Documentation* 70 (3): 443–460. https://doi.org/10.1108/JD-09-2012-0115.

Gnoli, Claudio, and Rick Szostak. 2014. "Universality Is Inescapable." In *Advances in Classification Research*. Seattle, WA: Ergon.

Gnoli, Claudio, and Roberto Poli. 2004. "Levels of Reality and Levels of Representation." *Knowledge Organization* 31 (3): 151–160.

Godfray, H. C. J. 2007. "Linnaeus in the Information Age." *Nature* 446 (7133): 259–260. https://doi.org/10.1038/446259a.

Godfray, H. Charles J. 2002. "Challenges for Taxonomy." *Nature* 417 (6884): 17–19. https://doi.org/10.1038/417017a.

Göhler, Gerhard. 2009. "'Power To' and 'Power Over.'" In *The SAGE Handbook of Power*, edited by Stewart Clegg and Mark Haugaard, 27–39. London: SAGE.

Goldman, Alvin, and Thomas Blanchard. 2016. "Social Epistemology." In *The Stanford Encyclopedia of Philosophy*, edited by Edward N. Zalta. Stanford, CA: Stanford University Metaphysics Research Lab. https://plato.stanford.edu/archives/win2016/entries/epistemology-social/.

Gordon, Dennis P. 2009. "Towards a Management Hierarchy (Classification) for the Catalogue of Life: Draft Discussion Document." Species 2000 & ITIS Catalogue of Life: 2009 Annual Checklist. http://www.catalogueoflife.org/col/info/hierarchy.

Grant, V. 2003. "Incongruence between Cladistic and Taxonomic Systems." *American Journal of Botany* 90 (9): 1263–1270. https://doi.org/10.3732/ajb.90.9.1263.

Green, Rebecca, and Giles Martin. 2013. "A Rosid Is a Rosid Is a Rosid . . . or Not." *Advances in Classification Research Online* 23 (1): 9–16. https://doi.org/10.7152/acro.v23i1.14228.

Greenblatt, Ellen, ed. 2010. *Serving LGBTIQ Library and Archives Users: Essays on Outreach, Service, Collections and Access*. Jefferson, NC: McFarland.

Groß, Anika, Cédric Pruski, and Erhard Rahm. 2016. "Evolution of Biomedical Ontologies and Mappings: Overview of Recent Approaches." *Computational and Structural Biotechnology Journal* 14 (August): 333–340. https://doi.org/10.1016/j.csbj.2016.08.002.

Guala, Gerald F. 2016. "The Importance of Species Name Synonyms in Literature Searches." *PLOS One* 11 (9): e0162648. https://doi.org/10.1371/journal.pone.0162648.

Guralnick, Robert, and Andrew Hill. 2009. "Biodiversity Informatics: Automated Approaches for Documenting Global Biodiversity Patterns and Processes." *Bioinformatics* 25 (4): 421–428. https://doi.org/10.1093/bioinformatics/btn659.

Hacking, Ian. 2007. "Kinds of People: Moving Targets." *Proceedings of the British Academy* 151: 285–318.

Haeckel, Ernst. 1866. *Generelle Morphologie Der Organismen. Allgemeine Grundzüge Der Organischen Formen-Wissenschaft, Mechanisch Begründet Durch Die von Charles Darwin Reformirte Descendenz-theorie*. Berlin: Druck und Verlag von Georg Reimer.

Hamer, Michelle, Janine Victor, and Gideon F. Smith. 2012. "Best Practice Guide for Compiling, Maintaining and Disseminating National Species Checklists." Global Biodiversity Information Facility. http://www.gbif.org/orc/?doc_id=4752.

Harding, Sandra G., ed. 2004. *The Feminist Standpoint Theory Reader: Intellectual and Political Controversies*. New York: Routledge.

Harvard T. H. Chan School of Public Health. 2020. "Coronavirus and Climate Change." Center for Climate, Health and the Global Environment | Harvard T.H. Chan School of Public Health (blog). https://www.hsph.harvard.edu/c-change/subtopics/coronavirus-and-climate-change/.

Hennig, Willi, D. Dwight Davis, and Rainer Zangerl. 1999. *Phylogenetic Systematics*. Urbana: University of Illinois Press.

Hinchliff, Cody E., Stephen A. Smith, James F. Allman, J. Gordon Burleigh, Ruchi Chaudhary, Lyndon M. Coghill, Keith A. Crandall, et al. 2015. "Synthesis of Phylogeny and Taxonomy into a Comprehensive Tree of Life." *Proceedings of the National Academy of Sciences* 112 (41): 12764–12769. https://doi.org/10.1073/pnas.1423041112.

Hine, Christine. 2008. *Systematics as Cyberscience: Computers, Change, and Continuity in Science*. Inside Technology. Cambridge, MA: MIT Press.

Hjørland, Birger. 2009. "Concept Theory." *Journal of the American Society for Information Science and Technology* 60 (8): 1519–1536. https://doi.org/10.1002/asi.21082.

Hjørland, Birger, and Hanne Albrechtsen. 1995. "Toward a New Horizon in Information Science: Domain-Analysis." *Journal of the American Society for Information Science* 46 (6): 400–425. https://doi.org/10.1002/(SICI)1097-4571(199507)46:6<400::AID-ASI2>3.0.CO;2-Y.

Hobern, Donald, Brigitte Baptiste, Kyle Copas, Robert Guralnick, Andrea Hahn, Edwin van Huis, Eun-Shik Kim, et al. 2019. "Connecting Data and Expertise: A New Alliance for

Biodiversity Knowledge." *Biodiversity Data Journal* 7 (August): e33679. https://doi.org/10.3897/BDJ.7.e33679.

Hodkinson, Trevor R., ed. 2011. *Climate Change, Ecology, and Systematics*. 1st ed. The Systematics Association Special Volume Series. Cambridge: Cambridge University Press.

Hopkins, G. W., and R. P. Freckleton. 2002. "Declines in the Numbers of Amateur and Professional Taxonomists: Implications for Conservation." *Animal Conservation* 5 (03): 245–249. https://doi.org/10.1017/S1367943002002299.

Hug, Laura A., Brett J. Baker, Karthik Anantharaman, Christopher T. Brown, Alexander J. Probst, Cindy J. Castelle, Cristina N. Butterfield, et al. 2016. "A New View of the Tree of Life." *Nature Microbiology* 1 (5): 16048. https://doi.org/10.1038/nmicrobiol.2016.48.

Huijbers, Chantal. 2020. "New Catalogue of Life Infrastructure Live." *COL* (blog). December 8. https://www.catalogueoflife.org/2020/12/18/infrastructure-live.

Hull, David L. 1988. *Science as a Process: An Evolutionary Account of the Social and Conceptual Development of Science*. Science and Its Conceptual Foundations. Chicago: University of Chicago Press.

Hull, David L. 2001. "The Role of Theories in Biological Systematics." *Studies in History and Philosophy of Science Part C: Studies in History and Philosophy of Biological and Biomedical Sciences* 32 (2): 221–238. https://doi.org/10.1016/S1369-8486(01)00006-1.

Hull, Matthew S. 2012. "Documents and Bureaucracy." *Annual Review of Anthropology* 41 (1): 251–267. https://doi.org/10.1146/annurev.anthro.012809.104953.

ICZN (International Commission on Zoological Nomenclature), ed. 1999. *International Code of Zoological Nomenclature*. 4th ed. London: International Trust for Zoological Nomenclature, c/o Natural History Museum. https://www.iczn.org/the-code/the-code-online/.

ICZN (International Commission on Zoological Nomenclature). 2018. "Can a Photograph or Holograph Be a Type Specimen?" https://www.iczn.org/outreach/faqs/.

Iles, Alastair. 2004. "Mapping Environmental Justice in Technology Flows: Computer Waste Impacts in Asia." *Global Environmental Politics* 4 (4): 76–107. https://doi.org/10.1162/glep.2004.4.4.76.

Index Fungorum. 2021. "Index Fungorum." Index Fungorum. http://www.indexfungorum.org/names/names.asp.

Institute of Medicine (US), and Health Sciences Policy Program (US), eds. 1999. *Toward Environmental Justice: Research, Education, and Health Policy Needs*. Washington, DC: National Academies Press.

International Association for Plant Taxonomy. 2011. "International Code of Nomenclature for Algae, Fungi, and Plants." International Association for Plant Taxonomy. July. http://www.iapt-taxon.org/nomen/main.php.

International Barcode of Life. 2015. "International Barcode of Life (IBOL)." http://ibol.org/.

ISKO Italia. 2004. "Integrative Levels Classification." International Society for Knowledge Organization, Italia. http://www.iskoi.org/ilc/.

ITIS (Integrated Taxonomic Information System). 2020. "ITIS." itis.gov.

IUCN (International Union for Conservation of Nature). 2019. "The IUCN Red List of Threatened Species." http://www.iucnredlist.org/.

IUCN (International Union for Conservation of Nature). 2020. "Who Uses The IUCN Red List?" IUCN Red List of Threatened Species. https://www.iucnredlist.org/about/faqs#Who%20uses%20the%20Red%20List.

Jacobsson, Bengt. 2005. "Standardization and Expert Knowledge." In *A World of Standards*, reprint, edited by Nils Brunsson, Bengt Jacobsson, and Associates. Oxford: Oxford University Press.

Jetz, Walter, Jana M. McPherson, and Robert P. Guralnick. 2012. "Integrating Biodiversity Distribution Knowledge: Toward a Global Map of Life." *Trends in Ecology & Evolution* 27 (3): 151–159. https://doi.org/10.1016/j.tree.2011.09.007.

Kate, K. T. 2002. "Global Genetic Resources: Science and the Convention on Biological Diversity." *Science* 295 (5564): 2371–2372. https://doi.org/10.1126/science.1070725.

Kellogg, Wendy A., and Anjali Mathur. 2003. "Environmental Justice and Information Technologies: Overcoming the Information-Access Paradox in Urban Communities." *Public Administration Review* 63 (5): 573–585. https://doi.org/10.1111/1540-6210.00321.

Kendig, Catherine, and Joeri Witteveen. 2020. "The History and Philosophy of Taxonomy as an Information Science." *History and Philosophy of the Life Sciences* 42 (3): 40. https://doi.org/10.1007/s40656-020-00337-8.

Kirk, Paul. 2017. Personal interview.

Krell, Frank-Thorsten, and Stephen A. Marshall. 2017. "New Species Described from Photographs: Yes? No? Sometimes? A Fierce Debate and a New Declaration of the ICZN." *Insect Systematics and Diversity* 1 (1): 3–19. https://doi.org/10.1093/isd/ixx004.

Kuhn, Thomas S. 1996. *The Structure of Scientific Revolutions*. 3rd ed. Chicago: University of Chicago Press.

Kunze, Thomas, Viktoras Didžiulis, and Yuri Roskov. 2013. "Proto-GSD in the Catalogue of Life a Case Study on Mollusca & Platyhelminthes." Presented at the 48th Annual European Marine Biology Symposium, National University of Ireland, Galway, August 19. http://134.213.156.20/sites/default/files/i4lifeposter_Galway_%20Kunze.pdf.

Lakoff, George. 2012. *Women, Fire, and Dangerous Things: What Categories Reveal about the Mind*. Chicago: University of Chicago Press.

Lakoff, George, and Mark Johnson. 2003. *Metaphors We Live By*. Chicago: University of Chicago Press.

Lakoff, George, and Mark Johnson. 2010. *Philosophy in the Flesh: The Embodied Mind and Its Challenge to Western Thought*. New York: Basic Books.

Landers, Jackson. 2016. "Big Data Just Got Bigger as IBM's Watson Meets the Encyclopedia of Life." Smithsonian, October 18. http://www.smithsonianmag.com/smithsonian-institution/ibms-watson-meets-encyclopedia-life-under-new-grant-180960772/.

Langridge, Derek Wilton. 1992. *Classification—Its Kinds, Elements, Systems, and Applications*. Topics in Library and Information Studies. London/Wagga Wagga, NSW: Bowker/Centre for Information Studies, Charles Sturt University.

Latour, Bruno. 2017. *Facing Gaia: Eight Lectures on the New Climatic Regime*. Translated by Catherine Porter. Cambridge, UK: Polity.

Lefebvre, Henri. 2011. *The Production of Space*. Translated by Donald Nicholson-Smith. Malden, MA: Blackwell.

Lefèvre, Wolfgang. 2001. "Natural or Artificial Systems? The Eighteenth-Century Controversy on Classification of Animals and Plants and its Philosophical Contexts." In *Between Leibniz, Newton, and Kant*, edited by Wolfgang Lefèvre. Dordrecht, Netherlands: Springer.

Leonelli, Sabina. 2016. *Data-Centric Biology: A Philosophical Study*. Chicago: University of Chicago Press.

The Library of Congress. 2020. "LC Linked Data Service: Authorities and Vocabularies (Library of Congress): Dogs." https://id.loc.gov/authorities/subjects/sh85038796.html.

Lima, Manuel. 2014. *The Book of Trees: Visualizing Branches of Knowledge*. New York: Princeton Architectural Press.

Littletree, Sandra, Miranda Belarde-Lewis, and Marisa Duarte. 2020. "Centering Relationality: A Conceptual Model to Advance Indigenous Knowledge Organization Practices." *Knowledge Organization* 47 (5): 410–426.

Liu, Hong-Mei, Li-Juan He, and Harald Schneider. 2014. "Towards the Natural Classification of Tectarioid Ferns: Confirming the Phylogenetic Relationships of Pleocnemia and Pteridrys (Eupolypods I)." *Journal of Systematics and Evolution* 52 (2): 161–174. https://doi.org/10.1111/jse.12073.

Locey, Kenneth J., and Jay T. Lennon. 2016. "Scaling Laws Predict Global Microbial Diversity." *Proceedings of the National Academy of Sciences*, May. https://doi.org/10.1073/pnas.1521291113.

Løvtrup, Soren. 1987. "Phylogenesis, Ontogenesis and Evolution." *Bolletino Di Zoologia* 54 (3): 199–208. https://doi.org/10.1080/11250008709355584.

Lowe, E. J. 2005. *The Four-Category Ontology*. Oxford: Oxford University Press. http://www.oxfordscholarship.com/view/10.1093/0199254397.001.0001/acprof-9780199254392.

Lowe, Ernest J. 2004. *The Possibility of Metaphysics: Substance, Identity, and Time*. Repr. Oxford: Clarendon Press.

Mai, Jens-Erik. 1999. "A Postmodern Theory of Knowledge Organization." *Proceedings of the ASIS Annual Meeting* 36:547–556.

Mai, Jens-Erik. 2011. "The Modernity of Classification." *Journal of Documentation* 67 (4): 710–730. https://doi.org/10.1108/00220411111145061.

"Map of Life." 2018. Map of Life. https://mol.org/.

Mathiesen, Kay. 2016. "Informational Justice: A Conceptual Framework for Social Justice in Library and Information Services." *Library Trends* 64 (2): 198–225. https://doi.org/10.1353/lib.2015.0044.

Matthias, Ryanne L. F. 2013. "What Is a Proto-GSD?" *Catalogue of Life* (blog). July 3. http://blog.catalogueoflife.org/2013/07/why-global-species-databases-and-what.html.

McGann, Jerome J. 2001. *Radiant Textuality: Literature after the World Wide Web*. New York: Palgrave.

McGinty, Maxine. 1969. "Batillipes Gilmartini, a New Marine Tardigrade from a California Beach." *Pacific Science* 23:394–396.

Mesibov, Bob. 2010. "Re: [Taxacom] GBIF: Perpetuating Probably Defunct Unpublished Names." *Taxacom—Biological Systematics Discussion List* (blog). May 24. http://taxacom.markmail.org/message/nnqgl4j2qnqxmpqf?q=perpetuating+probably+defunct+unpublished+order:date-forward&page=1#query:perpetuating%20probably%20defunct%20unpublished%20order%3Adate-forward+page:1+mid:ed3ewul7wd2xid6v+state:results.

Mesibov, Bob. 2018. "IPhylo: Guest Post: The Not Problem." *IPhylo* (blog). January 24. http://iphylo.blogspot.com/2018/01/guest-post-not-problem.html.

Mignolo, Walter. 2003. *The Darker Side of the Renaissance: Literacy, Territoriality, and Colonization*. 2nd ed. Ann Arbor: University of Michigan Press.

Mignolo, Walter. 2011. *The Darker Side of Western Modernity: Global Futures, Decolonial Options*. Latin America Otherwise: Languages, Empires, Nations. Durham, NC: Duke University Press.

Miksa, Francis L. 1998. *The DDC, the Universe of Knowledge, and the Post-Modern Library*. Albany, NY: Forest Press.

Mills, Charles W. 2011. *The Racial Contract*. Ithaca, NY: Cornell University Press.

Montenegro, María. 2019. "Subverting the Universality of Metadata Standards." *Journal of Documentation*, July. https://doi.org/10.1108/JD-08-2018-0124.

Montoya, Robert D., and Gregory H. Leazer. 2019. "Public Knowledge, Private Ignorance, and an Analytic of Knowledge Organization." *Proceedings from North American Symposium on Knowledge Organization* 7. https://doi.org/10.7152/nasko.v7i1.15563.

Montoya, Robert D., and Gregory H. Leazer. Forthcoming. "Subjects of Epistemic Oppression: Children and Young Adults as Underrepresented Library Communities in Kosovo and California." Manuscript Submitted for Publication.

Moore, Jason W. 2016. *Anthropocene or Capitalocene? Nature, History, and the Crisis of Capitalism*. Oakland, CA: PM Press.

Mullaney, Thomas S. 2012. *Coming to Terms with the Nation: Ethnic Classification in Modern China*. Berkeley: University of California Press.

Müller-Wille, Staffan. 2013. "Systems and How Linnaeus Looked at Them in Retrospect." *Annals of Science* 70 (3): 305–317. https://doi.org/10.1080/00033790.2013.783109.

Müller-Wille, Staffan, and Isabelle Charmantier. 2012. "Natural History and Information Overload: The Case of Linnaeus." *Studies in History and Philosophy of Science Part C: Studies in History and Philosophy of Biological and Biomedical Sciences* 43 (1): 4–15. https://doi.org/10.1016/j.shpsc.2011.10.021.

Nakata, Martin N. 2007. *Disciplining the Savages, Savaging the Disciplines.* Canberra, ACT: Aboriginal Studies Press.

Nicholson, Daniel J., and John Dupré, eds. 2018. *Everything Flows: Towards a Processual Philosophy of Biology.* 1st ed. Oxford: Oxford University Press.

Noble, Safiya Umoja. 2018. *Algorithms of Oppression: How Search Engines Reinforce Racism.* New York: New York University Press.

Nussbaum, Martha Craven. 2007. *Frontiers of Justice: Disability, Nationality, Species Membership.* The Tanner Lectures on Human Values. Cambridge, MA: Belknap Press of Harvard University Press.

Office of Parliamentary Counsel. 2020. "Declared Animal Policy under Section 10 (1)(b) of the Natural Resources Management Act 2004." Office of Parliamentary Counsel. South Australia.

Olson, Hope. 2002. *The Power to Name: Locating the Limits of Subject Representation in Libraries.* Dordrecht: Springer.

Orrell, Thomas M. 2016. Personal interview.

Ostrom, Elinor. 2009. "A General Framework for Analyzing Sustainability of Social-Ecological Systems." *Science* 325 (5939): 419–422. https://doi.org/10.1126/science.1172133.

Page, Roderic (@rdmpage). 2016a. ".@taxonbytes @timrobertson100 @aeolid Another Issue Is That @GBIF Is an Aggregation of Sources That Maybe Themselves Be Aggregations . . ." Twitter (blog). August 17. https://twitter.com/rdmpage/status/765812591683837952?ref_src=twsrc%5Etfw.

Page, Roderic (@rdmpage). 2016b. "@taxonbytes @timrobertson100 @aeolid Personally I'd like @GBIF to Take More 'Ownership' of Data Quality, but That's Politically Tricky." Twitter (blog). August 17. https://twitter.com/rdmpage/status/765813284297736197?ref_src=twsrc%5Etfw.

Page, Roderic (@rdmpage). 2016c. ".@taxonbytes @timrobertson100 @GBIF @aeolid And Some Data We Do Have Is Poor (e.g., @catalogueoflife Has Mangled Butterfly Names)." Twitter (blog). August 17. https://twitter.com/rdmpage/status/765810885390729220?ref_src=twsrc%5Etfw.

Page, Roderic (@rdmpage). 2016d. ".@taxonbytes @timrobertson100 @GBIF @aeolid . . . so Fixing 'at Source' Becomes Problematic." Twitter (blog). August 17. https://twitter.com/rdmpage/status/765812929614741504?ref_src=twsrc%5Etfw.

Page, Roderic (@rdmpage). 2018. "@taxonbytes, It's Only When You Start Trying to Assemble Backbones That You Realize How Messy, Parochial, and Internally Contradictory Taxonomic Classification Often Are. E.g. Lots of Previously Known Homonyms Have Come to Light Once People Started Thinking Globally." Twitter (blog). January 18. https://twitter.com/rdmpage/status/954100530099425285.

Papaioannou, Helena Iles. 2012. "Actually, Yes, It *Is* a Discovery If You Find Something in an Archive That No One Knew Was There." *The Atlantic*, June 21. https://www.theatlantic.com/technology/archive/2012/06/actually-yes-it-is-a-discovery-if-you-find-something-in-an-archive-that-no-one-knew-was-there/258812/.

Pape, Thomas, and F. Christian Thompson. 2017. "Systema Dipterorum." Systema Dipterorum. http://www.diptera.org/.

Parks, Lisa, and Nicole Starosielski, eds. 2015. *Signal Traffic: Critical Studies of Media Infrastructures.* The Geopolitics of Information. Urbana: University of Illinois Press.

Parr, Cynthia S., Bongshin Lee, Dana Campbell, and Benjamin B. Bederson. 2004. "Visualizations for Taxonomic and Phylogenetic Trees." *Bioinformatics* 20 (17): 2997–3004. https://doi.org/10.1093/bioinformatics/bth345.

Parr, Cynthia S., Robert Guralnick, Nico Cellinese, and Roderic D. M. Page. 2012. "Evolutionary Informatics: Unifying Knowledge about the Diversity of Life." *Trends in Ecology & Evolution, Ecological and Evolutionary Informatics* 27 (2): 94–103. https://doi.org/10.1016/j.tree.2011.11.001.

Patterson, David, David Remsen, William Marino, and Cathy Norton. 2006. "Taxonomic Indexing—Extending the Role of Taxonomy." *Systematic Biology* 55 (3): 367–373. https://doi.org/10.1080/10635150500541680.

Petrak, Franz. 1969. *Index of Fungi, 1920–39: List of New Species and Varieties of Fungi, New Combinations and Names Published.* London: Commonwealth Mycological Institute.

Pietsch, Todd. 2015. "Biodiversity: Methods and Goals of Systematics, Phenetics, and Cladistics." Presented at the Biology of Fishes (Course): Fish/Biol 311, University of Washington.

Pilsk, Suzanne C., Martin R. Kalfatovic, and Joel M. Richard. 2016. "Unlocking Index Animalium: From Paper Slips to Bytes and Bits." *ZooKeys* 550 (January): 153–171. https://doi.org/10.3897/zookeys.550.9673.

Pliny. 1472. *Naturalis Historia.* Venetiis: per Nicolaum Ienson.

Pyle, Richard. 2016. "Towards a Global Names Architecture: The Future of Indexing Scientific Names." *ZooKeys* 550 (January): 261–281. https://doi.org/10.3897/zookeys.550.10009.

Pyle, Richard L. 2008. "Names, Concepts, Codes and Lots of Confusion: An Introduction to Names of Taxonomic Organisms." Presented at the Biodiversity Information Standards (TDWG) Annual Conference 2008, Fremantle (Perth), Australia, October 19.

Queiroz, Kevin de, and Jacques Gauthier. 1992. "Phylogenetic Taxonomy." *Annual Review of Ecology and Systematics* 23 (January): 449–480.

Quine, W. V. 1989. *Quiddities: An Intermittently Philosophical Dictionary.* Cambridge, MA: Harvard University Press.

Raby, Megan. 2017. *American Tropics: The Caribbean Roots of Biodiversity Science.* Flows, Migrations, and Exchanges. Chapel Hill: University of North Carolina Press.

Rawls, John. 1999. *A Theory of Justice.* Rev. ed. Cambridge, MA: Belknap Press of Harvard University Press.

Rees, Tony. 2009. "Re: [Taxacom] Catalogue of Life (CoL) Management Classification Draft Document." *Taxacom—Biological Systematics Discussion List* (blog). July 21. http://taxacom

.markmail.org/search/?q=document#query:document+page:2+mid:vkb6h7kzh5qydnmq+state:results.

"Reflections on Suzanne Briet." 2017. In *Forthcoming in the Proceedings of the 11th French ISKO Colloquium*. Paris: ISTE Editions, London.

Reichhardt, Tony. 1999. "Catalogue of Life Could Become Reality." *Nature* 399 (6736): 519. https://doi.org/10.1038/21051.

Remsen, David. 2010. "Re: [Taxacom] GBIF: Perpetuating Probably Defunct Unpublished Names." *Taxacom—Biological Systematics Discussion List* (blog). May 2. http://taxacom.markmail.org/message/rx4bmoywcsb4g653?q=perpetuating+probably+defunct+unpublished.

Remsen, David. 2016. "The Use and Limits of Scientific Names in Biological Informatics." *ZooKeys* 550 (January): 207–223. https://doi.org/10.3897/zookeys.550.9546.

Reyes-García, Victoria, Álvaro Fernández-Llamazares, Maximilien Guèze, Ariadna Garcés, Miguel Mallo, Margarita Vila-Gómez, and Marina Vilaseca. 2016. "Local Indicators of Climate Change: The Potential Contribution of Local Knowledge to Climate Research." *WIREs Climate Change* 7 (1): 109–124. https://doi.org/10.1002/wcc.374.

Rheinberger, Hans-Jörg. 2010. *On Historicizing Epistemology: An Essay*. Cultural Memory in the Present. Stanford, CA: Stanford University Press.

Ribes, David, and Thomas Finholt. 2009. "The Long Now of Technology Infrastructure: Articulating Tensions in Development." *Journal of the Association for Information Systems* 10 (5). http://aisel.aisnet.org/jais/vol10/iss5/5.

Richards, Thomas. 1993. *The Imperial Archive: Knowledge and the Fantasy of Empire*. London: Verso.

Rinder, Lawrence, Howard N. Fox, Doug Harvey, and Tim Hawkinson. 2005. *Tim Hawkinson*. New York: Whitney Museum of American Art/Los Angeles County Museum of Art; distributed by Harry N. Abrams.

Robertson, Tim. 2016. Personal interview.

Robyn, Linda. 2002. "Indigenous Knowledge and Technology: Creating Environmental Justice in the Twenty-First Century." *American Indian Quarterly* 26 (2): 198–220.

Rogers, D. Christopher, Shane T. Ahyong, Christopher B. Boyko, and Cédric D'Udekem D'Acoz. 2017. "Images Are Not and Should Not Ever Be Type Specimens: A Rebuttal to Garraffoni & Freitas." *Zootaxa* 4269 (4): 455–459. https://doi.org/10.11646/zootaxa.4269.4.3.

Rorty, Richard. 2009. *Philosophy and the Mirror of Nature*. 30th anniv. ed. Princeton, NJ: Princeton University Press.

Rosch, Eleanor, Barbara Bloom Lloyd, and Social Science Research Council (US), eds. 1978. *Cognition and Categorization*. Hillsdale, NJ: L. Erlbaum; distributed by Halsted Press.

Roskov, Yuri. 2016a. Personal interview.

Roskov, Yuri. 2016b. Personal interview.

Roskov, Yuri. 2016c. Personal interview.

Roskov, Yuri. 2017. Personal interview.

Ruggiero, Michael A., Dennis P. Gordon, Thomas M. Orrell, Nicolas Bailly, Thierry Bourgoin, Richard C. Brusca, Thomas Cavalier-Smith, Michael D. Guiry, and Paul M. Kirk. 2015a. "A Higher Level Classification of All Living Organisms." *PLOS One* 10 (4): e0119248. https://doi.org/10.1371/journal.pone.0119248.

Ruggiero, Michael A., Dennis P. Gordon, Thomas M. Orrell, Nicolas Bailly, Thierry Bourgoin, Richard C. Brusca, Thomas Cavalier-Smith, Michael D. Guiry, and Paul M. Kirk. 2015b. "S1 Appendix. List of Sources Consulted for Proposed Higher Level Classification of All Living Organisms." *PLOS One* 10 (4): e0119248. https://doi.org/10.1371/journal.pone.0119248.

SæTher, Ole A. 1979. "Underlying Synapomorphies and Anagenetic Analysis." *Zoologica Scripta* 8 (1–4): 305–312. https://doi.org/10.1111/j.1463-6409.1979.tb00644.x.

Santos, Boaventura de Sousa, ed. 2008. *Another Knowledge Is Possible: Beyond Northern Epistemologies*. Reinventing Social Emancipation toward New Manifestos 3. London: Verso.

Santos, Boaventura de Sousa, and Vandana Shiva, eds. 2008. "Biodiversity, Intellectual Property, and Globalization." In *Another Knowledge Is Possible: Beyond Northern Epistemologies*. Reinventing Social Emancipation toward New Manifestos 3. London: Verso.

Savage, David. 2015. "Delta Smelt Legal Battle Heads to Supreme Court." *Los Angeles Times*, January 8. https://www.latimes.com/local/california/la-me-court-california-water-20150108-story.html.

Schalk, Peter. 2016a. Personal interview.

Schalk, Peter. 2016b. Personal interview.

Schlosberg, David. 2007. *Defining Environmental Justice: Theories, Movements, and Nature*. Oxford: Oxford University Press.

Scoville, Caleb. 2019. "Hydraulic Society and a 'Stupid Little Fish': Toward a Historical Ontology of Endangerment." *Theory and Society* 48 (1): 1–37. https://doi.org/10.1007/s11186-019-09339-3.

Seberg, O., G. Droege, K. Barker, J. A. Coddington, V. Funk, M. Gostel, G. Petersen, and P. P. Smith. 2016. "Global Genome Biodiversity Network: Saving a Blueprint of the Tree of Life—a Botanical Perspective." *Annals of Botany* 118 (3): 393–399. https://doi.org/10.1093/aob/mcw121.

Seddon, Jessica, and Ramesh Srinivasan. 2014. "Information and Ontologies: Challenges in Scaling Knowledge for Development." *Journal of the Association for Information Science and Technology* 65 (6): 1124–1133. https://doi.org/10.1002/asi.23000.

Sharma, Kriti. 2015. *Interdependence: Biology and Beyond*. Meaning Systems. New York: Fordham University Press.

Shera, Jesse H. 1965. *Libraries and the Organization of Knowledge*. 1st ed. Hamden, CT: Archon Books.

Shera, Jesse H. 1970. *Sociological Foundation of Librarianship*. New York: NY: Asia Publishing House.

Sloan, Phillip R. 1972. "John Locke, John Ray, and the Problem of the Natural System." *Journal of the History of Biology* 5 (1): 1–53.

Slota, Steve, and Geoffrey C. Bowker. 2015. "On the Value of 'Useless Data': Infrastructures, Biodiversity, and Policy." March. https://www.ideals.illinois.edu/handle/2142/73663.

Smiraglia, Richard P. 2001. *The Nature of "a Work": Implications for the Organization of Knowledge.* Lanham, MD: Scarecrow Press.

Smiraglia, Richard P. 2005. "Instantiation: Toward a Theory." *Proceedings of the Annual Conference of CAIS / Actes Du congrès Annuel De l'ACSI.* https://doi.org/10.29173/cais310.

Smith, Barry, and Werner Ceusters. 2010. "Ontological Realism: A Methodology for Coordinated Evolution of Scientific Ontologies." *Applied Ontology* 5 (3/4): 139–188. https://doi.org/10.3233/AO-2010-0079.

Smith, Bradley, ed. 2015. *The Dingo Debate: Origins, Behaviour and Conservation.* Clayton South, Australia: CSIRO Publishing.

Smith, Bradley P., and Carla A. Litchfield. 2009. "A Review of the Relationship between Indigenous Australians, Dingoes (*Canis dingo*) and Domestic Dogs (*Canis familiaris*)." *Anthrozoös* 22 (2): 111–128. https://doi.org/10.2752/175303709X434149.

Smith, Linda Tuhiwai. 2012. *Decolonizing Methodologies: Research and Indigenous Peoples.* 2nd ed. London: Zed Books.

Smithsonian Institution. 2017. "Global Genome Initiative." Smithsonian National Museum of Natural History. https://ggi.si.edu/.

Smithsonian Institution Bureau of American Ethnology. 1895. *Annual Report of the Bureau of American Ethnology to the Secretary of the Smithsonian Institution 44th.* Washington, DC: US Government Printing Office. http://archive.org/details/annualreportofbu48smithso.

Sneath, P. H. A., and Robert R. Sokal. 1973. *Numerical Taxonomy: The Principles and Practice of Numerical Classification.* A Series of Books in Biology. San Francisco: W. H. Freeman.

Sokal, Robert R. 1966. "Numerical Taxonomy." *Scientific American* 215 (6): 106–116.

Sontag, Susan. 2001. *On Photography.* New York: Picador USA.

Species 2000. 2014. "Catalogue of Life—Standard Dataset (Version 7)." Species 2000. September 23. http://www.catalogueoflife.org/sites/default/files/datafiles/2014_CoL_Standard_Dataset_v7_23Sep2014.pdf.

Species 2000. 2015a. "Catalogue of Life—About." Catalogue of Life. http://www.catalogueoflife.org/content/about.

Species 2000. 2015b. "Catalogue of Life—Source Databases." Catalogue of Life. http://www.catalogueoflife.org/col/info/databases.

Species 2000. 2015c. "Contributors." Catalogue of Life. http://catalogueoflife.org/content/contributors.

Species 2000. 2015d. “User Guide | Catalogue of Life.” Catalogue of Life. http://www.cata logueoflife.org/content/user-guide.

Species 2000. 2016a. “2016 Annual Checklist: Classification, Estimates & Extinct Taxa.” Catalogue of Life. http://www.catalogueoflife.org/annual-checklist/2016/info/hierarchy.

Species 2000. 2016b. “Catalogue of Life—2016 Annual Checklist: The 2016 Annual Checklist.” Catalogue of Life. http://www.catalogueoflife.org/annual-checklist/2016/info/ac.

Species 2000. 2016c. “Contributing Your Data | Catalogue of Life.” Catalogue of Life. http://catalogueoflife.org/content/contributing-your-data.

Species 2000. 2017a. “Catalogue of Life.” Catalogue of Life. http://www.catalogueoflife.org/.

Species 2000. 2017b. “Frequently Asked Questions | Catalogue of Life.” Catalogue of Life. http://catalogueoflife.org/content/frequently-asked-questions#5.

Species 2000. 2020. “2016 Annual Checklist: Taxonomic Tree.” Catalogue of Life. http://www.catalogueoflife.org/annual-checklist/2016/browse/tree/id/f2e91eb663bb9950c972474f65d2693b.

Species 2000. 2020 [2017]. *Catalogue of Life PLus (CoL+)*. Species 2000. https://github.com/CatalogueOfLife/general.

Species 2000. 2021a. “CatalogueOfLife/Coldp.” GitHub. https://github.com/CatalogueOfLife/coldp.

Species 2000. 2021b. “ChecklistBank.” Catalogue of Life. https://data.catalogueoflife.org/.

Species 2000. 2021c. “Metadata: Version 2020-12-01 of the COL Checklist.” https://www.catalogueoflife.org/data/metadata

Species 2000 Secretariat. 2015a. “Species 2000—About.” Species 2000. http://www.sp2000.org/about-0.

Species 2000 Secretariat. 2015b. “Species 2000—Home.” Species 2000. https://sp2000.org.

Srinivasan, Ramesh, Robin Boast, Jonathan Furner, and Katherine M. Becvar. 2009. “Digital Museums and Diverse Cultural Knowledges: Moving Past the Traditional Catalog.” *Information Society* 25 (4): 265–278. https://doi.org/10.1080/01972240903028714.

Srinivasan, Ramesh, and Jeffrey Huang. 2005. “Fluid Ontologies for Digital Museums.” *International Journal on Digital Libraries* 5 (3): 193–204. https://doi.org/10.1007/s00799-004-0105-9.

Srinivasan, Ramesh, Alberto Pepe, and Marko A. Rodriguez. 2009. “A Clustering-Based Semi-Automated Technique to Build Cultural Ontologies.” *Journal of the American Society for Information Science & Technology* 60 (3): 608–620.

Stengers, Isabelle, Bruno Latour, and Michael Chase. 2014. *Thinking with Whitehead a Free and Wild Creation of Concepts*. Cambridge, MA: Harvard University Press.

Sterner, Beckett, Joeri Witteveen, and Nico Franz. 2020. “Coordinating Dissent as an Alternative to Consensus Classification: Insights from Systematics for Bio-Ontologies.” *History and Philosophy of the Life Sciences* 42 (1): 8. https://doi.org/10.1007/s40656-020-0300-z.

Stevens, Hallam. 2013. *Life Out of Sequence: A Data-Driven History of Bioinformatics.* Chicago: University of Chicago Press.

Szostak, Rick. 2008. "Classification, Interdisciplinarity, and the Study of Science." *Journal of Documentation* 64 (3): 319–332. https://doi.org/10.1108/00220410810867551.

Szostak, Rick. 2015. "A Pluralistic Approach to the Philosophy of Classification." *Library Trends* 63 (3): 591–614. https://doi.org/10.1353/lib.2015.0007.

Tanselle, G. Thomas. 1980. "The Concept of 'Ideal Copy.'" *Studies in Bibliography* 33:18–53.

Tennis, Joseph T. 2002. "Subject Ontogeny: Subject Access through Time and the Dimensionality of Classification." *Challenges in Knowledge Representation and Organization for the 21st Century: Integration of Knowledge across Boundaries: Proceedings of the Seventh International ISKO Conference* 8:54–59.

Tennis, Joseph T. 2008. "Epistemology, Theory, and Methodology in Knowledge Organization: Toward a Classification, Metatheory, and Research Framework." *Knowledge Organization* 35 (3/2): 102–112.

Tennis, Joseph T. 2012. "The Strange Case of Eugenics: A Subject's Ontogeny in a Long-Lived Classification Scheme and the Question of Collocative Integrity." *Journal of the American Society for Information Science and Technology* 63 (7): 1350–1359. https://doi.org/10.1002/asi.22686.

Tennis, Joseph T. 2015. "Foundational, First-Order, and Second-Order Classification Theory." *Knowledge Organization* 42 (4): 244–249.

Thiele, Kevin, and David Yeates. 2002. "Tension Arises from Duality at the Heart of Taxonomy." *Nature* 419 (6905): 337. https://doi.org/10.1038/419337a.

Thomas, Claire. 2009. "Biodiversity Databases Spread, Prompting Unification Call." *Science* 324 (5935): 1632–1633. https://doi.org/10.1126/science.324_1632.

Thorpe, Stephen. 2009. "Re: [Taxacom] Catalogue of Life (CoL) Management Classification Draft Document—Stephen Thorpe—Edu.Ku.Nhm.Mailman.Taxacom—MarkMail." Taxacom. July 18. http://taxacom.markmail.org/search/?q=document#query:document+page:2+mid:oewytokyqj6f3xag+state:results.

"'Tree of Life' for 2.3 Million Species Released." 2015. Phys Org. September 19. https://phys.org/news/2015-09-tree-life-million-species.html.

Uetz, Peter. 2016. "The Reptile Database: General Information (and 'FAQ')." The Reptile Database. July 14. http://www.reptile-database.org/db-info/introduction.html.

United Nations. 1997. "UN Conference on Environment and Development." Earth Summit. May 23. https://www.un.org/en/conferences/environment/rio1992.

United Nations. 2017. "Agenda 21: Sustainable Development Knowledge Platform." United-Nations.Org. https://sustainabledevelopment.un.org/index.php?page=view&type=400&nr=23&menu=35.

Utteridge, Timothy. 2016. Personal interview.

Vanden Berghe, Edward, Gianpaolo Coro, Nicolas Bailly, Fabio Fiorellato, Caselyn Aldemita, Anton Ellenbroek, and Pasquale Pagano. 2015. "Retrieving Taxa Names from Large Biodiversity Data Collections Using a Flexible Matching Workflow." *Ecological Informatics* 28 (July): 29–41. https://doi.org/10.1016/j.ecoinf.2015.05.004.

Van der Hoorn, Berry. 2016. Personal interview.

Vaux, Felix, Steven A. Trewick, and Mary Morgan-Richards. 2016. "Lineages, Splits and Divergence Challenge Whether the Terms Anagenesis and Cladogenesis Are Necessary." *Biological Journal of the Linnean Society* 117 (2): 165–176. https://doi.org/10.1111/bij.12665.

Walker, Gordon P. 2012. *Environmental Justice: Concepts, Evidence and Politics.* London: Routledge.

Waterton, Claire, Rebecca Ellis, and Brian Wynne. 2013. *Barcoding Nature: Shifting Cultures of Taxonomy in an Age of Biodiversity Loss.* Milton Park, UK: Routledge.

Watson, Mark F., Christopher H. C. Lyal, and Colin Pendry, eds. 2015. *Descriptive Taxonomy: The Foundation of Biodiversity Research.* The Systematics Association Special Volume, vol. 84. Cambridge: Cambridge University Press.

Wheeler, Quentin D. 2004. "Taxonomic Triage and the Poverty of Phylogeny." *Philosophical Transactions of the Royal Society of London B: Biological Sciences* 359 (1444): 571–583. https://doi.org/10.1098/rstb.2003.1452.

Whitehead, Alfred North. 1920. *The Concept of Nature.* Cambridge: Cambridge University Press. http://archive.org/details/cu31924012068593.

Whitehead, Alfred North. 1938. Modes of Thought. New York: Free Press.

Whitehead, Alfred North. 1985. *Process and Reality: An Essay in Cosmology: (Gifford Lectures Delivered in the University of Edinburgh during the Session 1927–28).* New York: Free Press.

Whitehead, Alfred North. 1997. *Science and the Modern World: Lowell Lectures, 1925.* New York: Free Press.

Wiley, E. O., and Lieberman, Bruce S. 2011. *Phylogenetics: Theory and Practice of Phylogenetic Systematics.* 2nd ed. Hoboken, NJ: Wiley-Blackwell.

Wilkins, John S. 2011. "Philosophically Speaking, How Many Species Concepts Are There?" *Zootaxa* 2765 (1): 58–60. https://doi.org/10.11646/zootaxa.2765.1.5.

Wilson, Patrick. 1968. *Two Kinds of Power: An Essay on Bibliographical Control.* University of California Publications: Librarianship, vol. 5. Berkeley: University of California Press.

Wilson, Patrick. 1973. "Situational Relevance." *Information Storage and Retrieval* 9 (8): 457–471. https://doi.org/10.1016/0020-0271(73)90096-X.

Wilson, Patrick. 1977. *Public Knowledge, Private Ignorance: Toward a Library and Information Policy.* Contributions in Librarianship and Information Science, no. 10. Westport, CT: Greenwood Press.

Wilson, Patrick. 1983. *Second-Hand Knowledge: An Inquiry into Cognitive Authority.* Contributions in Librarianship and Information Science, no. 44. Westport, CT: Greenwood Press.

Winston, Judith E. 1999. *Describing Species: Practical Taxonomic Procedure for Biologists.* New York: Columbia University Press.

Witteveen, Joeri. 2015. "Naming and Contingency: The Type Method of Biological Taxonomy." *Biology & Philosophy* 30 (4): 569–586. https://doi.org/10.1007/s10539-014-9459-6.

Woodburn, Mathew. 2016. Personal interview.

Woodbury, Richard B., and Nathalie F. S. Woodbury. 1999. "The Rise and Fall of the Bureau of American Ethnology." *Journal of the Southwest* 41 (3): 283–296.

World Register of Marine Species. 2017. "WoRMS—World Register of Marine Species." http://www.marinespecies.org/sponsors.php.

Wulf, Andrea. 2015. *The Invention of Nature: Alexander von Humboldt's New World.* 1st US ed. New York: Vintage Books.

Youatt, Rafi. 2015a. *Counting Species: Biodiversity in Global Environmental Politics.* Minneapolis: University of Minnesota Press.

Youatt, Rafi. 2015b. *Counting Species: Biodiversity in Global Environmental Politics.* Minneapolis: University of Minnesota Press.

Young, Iris Marion. 2011. *Responsibility for Justice.* Oxford Political Philosophy. Oxford: Oxford University Press.

Young, Liam Cole. 2017. *List Cultures: Knowledge and Poetics from Mesopotamia to BuzzFeed.* Recursions: Theories of Media, Materiality, and Cultural Techniques. Amsterdam: Amsterdam University Press.

Zieleniec, Andrzej. 2018. "Lefebvre's Politics of Space: Planning the Urban as Oeuvre." *Urban Planning* 3 (3): 5–15. https://doi.org/10.17645/up.v3i3.1343.

Zimmer, Carl. 2016. "Scientists Unveil New 'Tree of Life.'" *New York Times,* April 11. http://www.nytimes.com/2016/04/12/science/scientists-unveil-new-tree-of-life.html?action=click&contentCollection=Science&module=RelatedCoverage®ion=EndOfArticle&pgtype=article.

Index